What My Friends Say

"I have known Charles "Chuck" Andrews on both a professional and personal level for over four decades. In the late 1970s, I was a police officer for the City of Lake Jackson, and he was a law enforcement explorer scout. During our time together, Chuck rode with me for hundreds of hours patiently listening to my training, guidance, and many stories. On several occasions we found ourselves on calls that were quite dangerous and required Chuck to cover my back as well as care for victims and community citizens. And, as was his nature, he did that with the utmost compassion, efficiency, and professionalism.

As time moved on, Chuck's career blossomed in law enforcement and security, and I transitioned from police work to community building as an inter-governmental liaison, involved with professional speaking, training, and consulting throughout the United States and Canada. Over the years we have maintained a close, personal relationship, and many times I have called upon Chuck's advice, expertise, and insight into multiple projects. He is an amazing person and an incredible professional in his field."

– **Clint Hackney**
CEO/Former Police Officer

"Chuck's in-depth knowledge of the security industry is unparalleled. He navigated the transition from policing to private security better than most. Chuck knows everybody in the business and is highly respected. I will be the first to buy his book."

— **Ed Davis**
Founder of Edward Davis Company,
Former Boston PD Commissioner

"Since the age of thirteen, as a law enforcement Explorer, Chuck's mission has been to serve and protect his community. Over the years, his community and those he serves and protects have significantly grown. Chuck's contributions as a police officer, businessman, teacher, and technologist are far-reaching and impactful. Chuck's network of talented people from various backgrounds with different areas of expertise, combined with sharing that network, has not only grown his influence but has also made a difference."

— **Mark Sullivan**
Former Director,
United States Secret Service

"Over the fifteen years I've known Chuck, he's been my boss, mentor, colleague, and more importantly, my friend. He's a respected leader and influencer in our industry, and he's mastered the art of building relationships! I was there when the idea of Friends of Chuck (FOC) came about, and he's taken that idea and run with it. He's one of the hardest-working people I know, and I personally know dozens of people he has helped professionally. I, too, have been a direct recipient of that help, so I was honored and humbled to have been asked to write this testimonial!"

— **JC**
Director of Security, Certified Fraud Examiner,
Certified Forensic Interviewer, Master of Security
Management, and FOC Member/Beneficiary

"If you want to know about the security industry, networking, and building relationships then you will want to read this book. As someone who has known the author for years and watched him in action whether helping friends find new jobs to connecting people, companies, and ideas. Chuck is a literal tour de force, and this book will illustrate his reach and what one passionate person can do to unite an industry and community."

– **FHW**
U.S. Special Forces,
United States Marshals Service

"Chuck is the ultimate influencer and networker in the security industry. I have known Chuck for many years and respect his security expertise, leadership acumen, and ability to link people to enhance their business strategies and goals. Chuck and I have worked together, and he has definitely made a positive impact on my professional career. This book will definitely show the reader how to become an effective influencer and understand the force multiplier that is known as networking! As the successful creator of "Friends of Chuck" (FOC), Chuck knows how to create and brand a business and make it truly effective in marketing a business. Well worth reading!"

– **Mike Howard**
Former Chief Security Officer – Microsoft Corporation
and author of *The Art of Ronin Leadership*
and *Executing the Art of Ronin Leadership*

"Chuck Andrews – police officer, ex-Police Explorer Post 79 advisor, security expert, and hardworking, dedicated, influential, inspirational, funny, faithful, family man, friend, and mentor. Plus, let's not forget—a Texan. Forever FOC!"

– **Sergeant Kevin Jamison**

"I have been a friend and colleague of Chuck's for over ten years and a charter member of "Friends of Chuck" (FOC). I am excited to endorse Chuck's book regarding the art of influencing relationships in the security industry. When I first met Chuck, we hit it off almost immediately; his friendship, collegial attitude, advice, and guidance are unparalleled and have helped me significantly as a single shingle consultant. It can be challenging when you are on your own. Having a friend and colleague to turn to as a sounding board is invaluable. I am thrilled that Chuck will put this formula into a book and share it with others."

– **Jeffrey A. Slotnick, CPP, PSP**
President Setracon Incorporated
ASIS International North American Board of Directors

"Knowing Chuck for nearly two decades, he is indeed larger than life in many facets! He and his knack for opportunity and his "connection with people" are unmatched… And, I am honored to be included in the many benefits of FOC."

– **Darin Dillon**
Sr. Director Energy LenelS2,
Board Member – Energy Security Council

"Chuck Andrews is a pioneer and trailblazer like no other… A maverick and enigma within the security industry, with a larger-than-life character. His gift of connecting and nurturing relationships for the good and benefit of all, is second to none. I'm humbled to be able to walk alongside Chuck, as a valued friend in pure admiration for all he does."

– **Lee Oughton**
Co-founder of The Kindness Games

"My transition from the Combat Control Teams was more difficult than expected. Combat Controllers are the most lethal human weapon system in the Department of Defense and due to our warriors' ethos of "quiet professionals," most of America has never heard of us; thus, this puts us at a tremendous disadvantage when applying for jobs. I am lucky and very thankful to have been introduced to Chuck. I've told many friends and colleagues "Chuck knows everyone. Chuck probably knows Jesus." His connections and network helped me start a career in the corporate security industry and definitely eased my transition into civilian life from the teams."

– **Eric Hohman**
Combat Controller,
United States Air Force

"I wholeheartedly endorse *Yes, S.I.R.* by one of the singular icons of the security profession. Chuck Andrews delivers a template of knowledge for current and future generations of individuals in our industry, exemplifying the profession's movement from the boiler room to the boardroom. His life, observations, and recommendations in this book represent "the best and the brightest" from an exceptional colleague who practices that qualitative excellence in which he excels…while equally important, challenging and educating others to excel beyond their expectations."

– **Ray Humphrey**
Past President, ASIS,
International and co-founder,
Past President, International Security
Management Association (ISMA)

"I have known Chuck for many years through the Security industry. His intelligence, wit, energy, passion and his ability to connect with people is quite amazing. Chuck is the essence of a genuine person and professional. He has made a mark in the industry and influenced huge numbers of people while also showcasing innovation in moving the industry forward. His positivity and spirit is a treat for anyone working with him in any capacity."

– **Bonnie Michelman, CPP, CHPA**
Executive Director of Police and Security
Mass General Hospital and Mass General Brigham Corporation
Past President; ASIS International, International Association
for Healthcare Security and Safety (IAHSS) and International
Security Management Association (ISMA)

YES, S.I.R.

YES, S.I.R.

The Security Influencer's Guide to success using
Strategy, **I**ntelligence, and **R**elationships

CHUCK ANDREWS, CPP
#1 Global Security Influencer

Copyright © 2022 by Chuck Andrews, CPP

Softcover ISBN: 979-8-88739-004-8
eBook ISBN: 979-8-88739-005-5

All rights reserved. No part of this book may be reproduced or transmitted in any form or by any means, electronic or mechanical, including photocopying, recording or by any information storage and retrieval system, without permission in writing from the copyright owner. For information on distribution rights, royalties, derivative works or licensing opportunities on behalf of this content or work, please contact the publisher at the address below.

Printed in the United States of America.

Cover Design: Heidi Caperton

Although the author and publisher have made every effort to ensure that the information and advice in this book was correct and accurate at press time, the author and publisher do not assume and hereby disclaim any liability to any party for any loss, damage, or disruption caused from acting upon the information in this book or by errors or omissions, whether such errors or omissions result from negligence, accident, or any other cause.

Friends of Chuck
117 State Hwy 332 West Suite J
PO Box 210
Lake Jackson, TX 77566

Dedication

There are many people—friends, colleagues, cops, spies, family members, criminals—events, moments, organizations, hospitals, companies, and more to thank in respect to the publication of this book. My life is very intentional, and I've lived it in a manner of RENAISSANCE in all that I do. If a life is worth living, then I say it's worth living well!

Table of Contents

Foreword . . . . *xv*

1. "Junior Pig" 1
2. 4,000 Hours and Then Some 7
3. A Dark Truck at 2 AM 13
4. Yes, S.I.R. and Building My Brand 19
5. Friends of Chuck (FOC) 35
6. Yes, S.I.R. 41
7. Dream Big 45
8. Have Fun 53
9. Get Stuff Done 61
10. Strategy 69
11. Intelligence 77
12. Relationships 85

Join Friends of Chuck! *89*
About Chuck *91*
What My Friends Say (cont.) *95*

Foreword

Dear Reader,

I am an ugly crier. If you were with me right now as I am writing this, you would see tears well in my eyes. Little drops of gratitude and pride. So thank goodness for both of us that this is a written format. It is a huge honor to write the foreword of this book.

I am Chuck's youngest daughter, and I couldn't be prouder of this huge milestone he has accomplished. Not only because of the immense amount of work that went into writing this book, but because of how many people I know it is going to help. I should know; I've lucked out by having him help me my whole life. I'm so excited that you get to learn his tips and tricks to networking and building a wildly successful career.

The last 30 years, I've been privileged to watch all the lives he has touched, all the relationships he's built and businesses he has helped thrive. Seeing him put his methodology into words and share with you is something I know will change the way you see building your career and meeting new people.

My dad is nothing short of influential. He brings together the right people with the optimal strategy to make a win-win-win situation for all those involved (yes, triple win). If you are wanting to increase your realm of influence and be more strategic in your relationships and your career, rest assured that you are in the right place.

YES, S.I.R.

There is nothing better than the S.I.R. methodology to learn how to master relationships and thrive in your career. I had the ultimate cheat code in building businesses and relationships—I grew up as the ultimate Friend of Chuck's insider. I think you will find this book a cheat code as well.

Happy Reading,

Lauren Andrews
FOC Super Fan & Youngest Daughter

While Chuck served as a Police Explorer in Clute, Texas, in the late 1970s, I enjoyed watching him work and become interested in being a law enforcement and security officer. He was always there when needed and learned things quickly. In fact, he picked up on the job faster than anyone I've ever seen in over 50 years in the law enforcement business. So, when the opportunity came to hire a police officer for the summer, Chuck was the first guy I considered. He came to the job with high energy, aimed at getting the job done! He brought such unbelievable energy to the office that things happened when Chuck was around. He had the tenacity to find things, not just as a policeman, but every day he surprised me, even as a rookie. He worked on cases and followed the leads to make things happen. When his time was over, Chuck moved on to bigger things he was destined to do. He saw his calling in security, teaching classes so others could protect themselves, crime prevention, and community policing. Chuck never ceased to amaze me with his knowledge, empathy, and other gifts he shares with folks. It's my privilege to now call Chuck a colleague and friend. I treasure that more than anything. Enjoy the book. You will learn from it and maybe it will help you enhance your career as well.

Chief Glenn Patton (Ret.)

"Junior Pig"

If I've learned one fact about the security industry, it's this: it's all about relationships!

And this is something I began learning from an early age, especially from my stepdad. You see, my parents divorced when I was young, and after my mother remarried, I lived with my mom and stepdad. At the same time, my real dad was also involved in my life, so I guess you could say I was fortunate enough to have two dads influencing me (It also meant I had double the expectations!)

But before I talk further about my family, let's start at the beginning. I was born on September 11 (a key date in my life for many reasons) in Houston and raised in Lake Jackson, Texas, some fifty miles to the south on the Gulf Coast. Lake Jackson is home to Dow Chemical's Freeport facility, the largest chemical plant in the world, making it a unique place in which to grow up.

YES, S.I.R.

Dow drew top talent from around the world, many of them graduates from Massachusetts Institute of Technology (MIT), Harvard, Texas A&M, etc. So, from a young age, we all saw how a good company could draw world-class talent. In Lake Jackson, it didn't matter if you had money because we were all Dow kids, and everyone knew how to weld or use duct tape to fix just about anything. We were by-products of a great work ethic. Skin color, gender, and rich or poor didn't matter. To us, we were all Texans.

My first introduction to the notion of security and safety was probably from Dow because the one topic the company constantly repeated was safety, and everyone who worked there brought that home. Every house had a fire extinguisher in the kitchen and Tot Finder stickers on bedroom windows. Dow once ran a huge safety campaign with the slogan "Life is Fragile, Handle with Care."

Getting back to my family, I can see now that my stepdad and my biological dad were a study in contrasts.

My biological dad, Larry, was a MENSA member and also an electrical engineer. He was brilliant and hard-working but not a social guy. When he was just fifteen, he bought his own car and fixed it up but he didn't tell my grandma. His dad—my Granddaddy Lou—was part of the Andrews dynasty, most of them in the oil, gas, and chemical operations. Dad thought college was a joke—it was too easy—but he went through it anyway and made straight As. He started out working for Dow and morphed to be the number one project manager in the world, later forming his own successful company.

In contrast, my stepdad, Bill, was hillbilly-born, raised in Arkansas, and studied mechanical engineering at the University of Arkansas. Bill could literally make and fix anything, whether it was made of metal or wood, from boats to houses, you name it.

"JUNIOR PIG"

He and I renovated four boats over the years, and he also remodeled our houses. He taught me hunting, fishing, and everything outdoors.

My stepdad was as social as my dad was non-social. He remembered everybody's names. His retirement inspired a huge party, and his death inspired an even bigger funeral, with over 500 attendees. As the emcee, at one point during the service I told everyone I wanted to call my dad up, which confused them because they assumed I was talking about my stepdad. My biological dad was quite moved by my invitation, but he rose to the occasion and said my stepdad had raised a son to be proud of. People were deeply impacted by his remarks. Having my dad speak at my stepdad's funeral broke all the rules, which is something that needs to be broken in more families.

My stepdad worked for Dow in the magnesium cells, the toughest place to work at the company. He didn't rise up the ranks but preferred to stay where he was, taking care of his crew, because they were more important to him than status or advancement. My stepdad taught me the importance of not only knowing the right people, but taking the time to truly know them. His friendships stretched back decades, and his ability to recall everyone's names and details reinforced to me that relationships are critical. I wish I could say I always followed his example, but it wasn't until my forties that I really started to get what my stepdad had been showing me over the years.

My entire family—my real dad had remarried as well—all got along and worked together, as evidenced by my birth dad's willingness to speak at my stepdad's funeral. Another evidence of this cooperation happened when I was in the seventh grade. One day, my dad called my mom and told her he wanted me to spend an entire year with him and that side of the family up in Alaska where they had just moved.

YES, S.I.R.

Mom said, "You bet. I'll tell him when he gets home." I was two days into my seventh-grade year, but off to Alaska I went to spend an entire year with my dad, step-mom, and stepsisters. What an adventure that was!

Law Enforcement Explorers

Two events happened in seventh grade that changed my life: spending the year in Alaska and becoming a "junior pig" by joining the Law Enforcement Police Explorers.

My stint with the Law Enforcement Police Explorers all started one day in 1976 when I was out in my front yard practicing karate in my kung fu outfit (think orange belt, kung fu pants, and no shirt). My buddy Pete was riding by on his Schwinn bike and suddenly skidded to a stop.

"Where you headed?" I asked.

"To the police station," he said.

"Why? Isn't that for criminals?"

He then told me that no, he was going to the Law Enforcement Police Explorers weekly meeting where he would get to ride around in a police car and work in the dispatch center. Presented with the opportunity to do real-life police stuff, I quickly decided I wanted in.

I ran and told my mom about it, but she, too, thought this program must be for juvenile delinquents. After I corrected her misconception, while begging, she said okay, and soon she was driving me to the police station in our family's wood-paneled station wagon.

By the time the meeting was over, I had discovered my life's calling, and my whole life was mapped out. Like a lot of kids, I grew up watching all the great cop shows: *T.J. Hooker*, *CHiPs*, *Adam-12*, and so on. I also loved shooting guns and playing cops

"JUNIOR PIG"

and robbers. So, when the opportunity came to ride around with a cop and protect and serve, I jumped.

Eventually, I learned that the Law Enforcement Police Explorers were called Law Enforcement Exploring, a division of the Boy Scouts of America since the early 1950s. (I also found out that our local police Explorer post had its own unmarked 1976 Explorer Scout patrol car!) Being a "junior pig" meant I was no longer popular in school, but I didn't care: I got to ride in a cop car, wear a police belt, and make the community safe, all of which I did for thousands of hours.

As Law Enforcement Explorers, we performed community service projects, road patrol, helped with crime prevention, and even got to work in jail operations. We had a level of formality and ranks, and we met weekly in the courtroom with our police advisor. I couldn't wait to get my uniform and saved up $200 to buy my own bullet-proof vest from Safariland, which I still have today. We also had some crazy fun. Once we found some discarded mannequins behind the Brazos Mall and took them out to an old wrecker yard to do "ballistic tests" on them.

As a Police Explorer, I noticed certain officers handled people better than others. The saying that you gather more bees with honey certainly applies to law enforcement. I learned that, even when someone was being a total asshole, you could almost always win them over if you killed them with kindness. But, there are exceptions!

When in law enforcement, you're in the people business. From how you gesture, to how you speak, to how you approach others, it's all about how you treat people. For instance, I learned to get down on one knee when talking to a young child so as not to intimidate them. Basically, I developed the ability to become a chameleon with all different types of people in order to get them to relate to me and be more cooperative. As a result of using this chameleon effect, people were drawn to me because I offered them "honey."

They would call on me first at the department because I treated people with respect and trust. As an example, a horrific crime occurred and the person wouldn't talk to any other officer but me. That was a direct result of being in the people business and doing it right. Now, after learning how to be a chameleon, I find that I'm less judgmental and less likely to throw the first stone.

In law enforcement, you also learn to think two or three steps (shots) ahead, which was a lesson my stepdad taught me from his time living above a pool hall and helping pay his way through college playing eight-ball. This habit of thinking ahead even spilled over into my schoolwork because I took my SAT test in ninth grade, rather than in the twelfth grade.

At a young age, I chose a life of service in law enforcement. I knew it wouldn't pay a whole lot, but I chose to serve over making money—service is what I feel we're on this planet to do. I've never regretted my decision because becoming a "junior pig" has led me to where I am today.

4,000 Hours and Then Some

I was only thirteen when I started going on patrol with the Police Explorers. Despite my youth and inexperience, I got thrown into a lot of real-life situations, and I learned fast.

When out on patrol, we'd get a raw feed from society. We'd roll up on a scene after driving 85 mph with lights and sirens in a 35-mph zone to answer a call, never sure exactly what we'd find. Every day, every call was different. Would we save a life? Lose our lives? Make a difference in somebody's life? Our day never had any agenda, which was a real intrigue and challenge for me. We had to be extremely resourceful and vigilant, for our safety and that of others.

As time went on, I created a unique, 360-degree awareness, and my listening became more powerful. More powerful still was my increased awareness of others' body movements and gestures. I would assess all this information before anyone even spoke.

YES, S.I.R.

The first time I saved someone's life happened one weekend. We rolled up first on the scene of a head-on collision, the vehicle's engine already on fire. The officer I rode with took one side of the car, and I took the other. Now, I was a pretty big kid, so I could handle whatever came. The engine started to burn through the firewall, and the victim was unconscious on the front seat, unbelted. We had to act. Fast. Thanks to my dad, I knew not to break out the windows because the incoming air would feed the fire. Instead, we emptied the extinguisher onto the fire and then used a flashlight to break open the windows to unlock both doors and drag the victim out—I can still hear the sounds of the burning wreckage today.

That victim survived, and I ended up encountering the same situation some twenty or thirty times throughout my career, each time having to weigh whether to move the victim or not. But that first experience let me know I could actually make a difference and save someone's life.

The Clint Era

The officer I rode with the most as a Law Enforcement Explorer was Officer Clint Hackney, whom I met after I had moved police Explorer Posts, from #351 to #396, Clute to Lake Jackson Police Department. We hit it off from the start. He was eager to teach, and I was eager to learn.

Clint lived nearby, so I often went to his house to endure one-on-one training. I also got my brown belt in kung fu karate from him. I was as big as he was if not bigger, but he'd been an all-American football player in high school and had also played for Southern Methodist University, so he had developed the discipline that comes from being an athlete. When I met him, I was sixteen, and he was probably twenty-five. I had everything but a handgun on my Sam Brown belt, although I did have the Remington 870 (12 gauge-shotgun) in the patrol car. He'd have me grab the shotgun when we got to scenes that warranted a level of protection!

Clint knew a lot about everything, but what he knew best was people. He taught me how to read eyes, body language, and voice modulation, as well as how to "clear" a room. Because he could read people, he knew how and when to take someone down so we didn't end up dead. After a while, it got so I didn't have to talk with him to know what needed to be done on a scene. What started as a student/teacher relationship morphed into a full-on partnership, and we later became colleagues.

One evening, we were at Mr. Gatti's Pizza, eating a pizza on the hood of the patrol car so we could be ready to roll at a moment's notice. Sure enough, we got a domestic violence call. The dispatcher's voice was distraught from what was going on, and when we arrived, we heard screaming and yelling. The house was in a tough part of town on a street I grew up on. I also knew the father who lived there was a total piece of work.

After we arrived, Clint jumped out the door, assessing the situation with his Model 66 357 Magnum at the front door. All the while, I kept the dispatcher apprised as to what was going on, so the information flow would get documented. I said something like, "417 approaching door at gunpoint, woman screaming, children screaming, stand by for potential forced entry." By this time, I'd situated myself behind the engine block of the car for cover. Now, there's a difference between cover and concealment. Cover is an engine block. Concealment is just a bush.

I next took one of the spotlights on my side of the car and used it as a diversion to light the house but not Clint. All the while, Clint was kicking the front door in. After he made entry, I was on the perimeter, and I pointed the shotgun toward the front of the house. I heard what I thought was gunfire, but I couldn't tell who or what, so I radioed it in, along with the fact that I would soon be without my radio—not having a portable one on me.

After Clint gained entry, still amid all the screaming and yelling, he kung fu'd the suspect and handcuffed him. And then, as is pretty typical, the daughter—who was by now covered in blood—crept up from behind to attack Clint by starting to jump on his back, all because Clint was arresting her dad. Thinking fast, I grabbed her and pulled her off.

Clint had been pretty focused on arresting the man, and he trusted me enough to know I had his back. As I guarded Clint, I told her—and this was something else I learned from Clint—"Miss Y, Paul [her brother] and I are friends." I was trying to be the chameleon, trying to find the common denominator with her in order to defuse the situation.

The man and his daughter were both arrested for domestic violence, although I have no idea how it all played out because we Explorers never went to court. It was fun, but it was real—no plastic badges here.

Dispatching

During my senior year in high school, one of the full-time dispatchers went on maternity leave, so the department hired me to fill in. By that time, I had racked up at least 3,000 hours on patrol as a Law Enforcement Explorer. That, plus my maturity and demonstrated competency, had gotten me the job, unsolicited.

Now, hiring a kid in his late teens as a police dispatcher was—and still is—unprecedented. But they trusted me totally. I worked graveyards and weekends. Dispatching often doesn't get the respect it deserves. It is a very difficult job—the definition of multi-tasking—and all police officers should spend time dispatching to give them a better understanding of their own jobs.

Milestones

After high school, I started college at Sam Houston State University in criminal justice. I was still a Law Enforcement Explorer and still dispatching. At that time, the department began entrusting me to do more: write tickets and police reports, sometimes even type up a warrant. I took on as much as possible that an officer would have to perform as part of their job. I even read every page of the Texas Penal Code and the Code of Criminal Procedure.

Over the summer, I had gone to the police academy at the junior college. With my combined Police Explorer experience and the college credits I already had, I obtained my police commission. Chief Glenn Patton, who had known me as a Police Explorer, gave me my first real job at the Richwood Police Department, working graveyards. When I started, I made $8.64/hour. I wasn't even old enough to buy my own bullets—something you had be twenty-one to do!

Typically, new police hires have to spend six months on patrol and demonstrate a host of competencies before they're allowed to go solo, called the FTO program. Not me. My time on probationary patrol was only two days because of the many hours I had logged and the many competencies I had already demonstrated as a Police Explorer. Being a "junior pig" had paid off big time, and attending and testing at the academy became an informality for me.

My first traffic stop working for Richwood was in the fog. I had pulled over a 1977 Ford Thunderbird driven by a guy with one leg. I didn't write the ticket because I felt bad for him, but I probably hold the record for tickets written in that town. Back then, State Highway 288 ran through our town, carrying all of Houston with it. In fact, one year I wrote more tickets than anybody—it was as if half of Houston drove through our town that summer.

YES, S.I.R.

Ours was a one-car town, so whenever I was on duty, it was all on me—a lot of responsibility. I was also protective of my community. I once got in trouble with Chief Patton because I told a lady on a traffic stop that "You don't speed in my town."

Thanks to Police Explorers, I was pretty much ready for whatever came at me. And, one night, that's just what happened.

A Dark Truck at 2 AM

The time I spent in Police Explorers and the training I had under Clint all came into play one weekend.

It was two o'clock in the morning, and I was all by myself on patrol working for the Richwood Police Department. I was only twenty years old, but my chief let me come home on the weekends to work on patrol while I was getting my degree as a criminal justice major at Sam Houston State University. That night, I noticed a truck traveling northbound toward Houston. When I got up behind this truck, I saw that his license plate light was out. It was an older, beat-up truck, but it hauled some nice equipment, which was unusual.

I couldn't see the driver, but because of his busted light I had probable cause to pull the truck over at an area of town known as Three Bridges. He stopped right before the first bridge, and I called in the truck's plate number to dispatch and then got out of my vehicle. From the time I first saw his truck, I had been looking at fifty different things, totally aware of everything I could be before I pulled him over, trying to assess who he was and what the threat might be. I performed this threat assessment thanks to my training from Clint and the Police Explorers and to my sixth sense.

After I got out of my patrol car, I approached the truck, staying as close to it as possible, and employed a maneuver we called "cutting the pie." As you approach an unknown vehicle, you go through green, yellow, and red zones as you "cut the pie" and expose yourself to more risk.

Again, given the early morning hour and odd cargo, something didn't seem right. I lit up his driver's side mirror with my flashlight, which was also a tactical move, and told him to place his hands on the steering wheel, but he refused to comply. I asked to see his hands because those two hands, those ten digits, are what can kill you, and if you can control them, you're 95 percent of the way to safety.

As I walked up, I again asked to see his hands, but he again didn't acknowledge me, which was a huge red flag. At this time, he couldn't see me because I'd blinded him with my flashlight, holding the light away from my body so he wouldn't know exactly where I was—another tactic Clint taught me.

By now, my instincts—probably the most important element to preserving my safety—told me this guy was bad news. I had taken one quick look at him, and, as the saying goes, "If it looks like a duck, walks like a duck, and there are feathers coming out the window, it's probably a duck." Despite my repeated requests, he still didn't respond. I banged on the side of the truck as I approached and told him this was his last warning. I elevated my language and said something like, "Put your f-ing hands on the steering wheel! I'm not telling you a fourth time!"

My flashlight caused enough reflection to light up the interior of his cab so I could see he had his hands deep in his lap, which was another problem. I got my gun out at the ready, not

A DARK TRUCK AT 2 AM

yet pointed toward him, but down at my side. Again, I had to get control of his hands. By this point, I had already confirmed there was nobody else in the vehicle. Finally, he slowly started to raise his hands, and, as he did, I could see the reflection of something chrome in his crotch area—a locked and cocked 380!

I aimed my gun at him while keeping at a distance so he couldn't grab my weapon. To accomplish this, I stood in the highway, perpendicular to his truck. In a shooting situation, you don't want to aim at somebody's head because they can move it very quickly. Instead, you want to aim at body mass. His window was partly open, and, as I aimed at him, I said, "I see your gun. This will be a really bad choice for you because I *will* kill you." Using my modulation, serious tone, loud voice, and abundant strong language, I communicated to the guy that I was serious. By that time, my finger was on the trigger, and I was prepared to shoot him if he tried to grab the gun and shoot me.

Everything I did was the culmination from that moment all the way back to my Police Explorer Training time, such as the 21-foot rule we learned in the Police Academy. That rule means that a guy with a knife who is twenty-one feet away can stab you before you have time to get your gun out. Clint taught me to create a barrier and get a suspect's hands away from anything that could harm me. So, when this guy put his hands up, I told him to stick them outside the window. I was still blinding him with my light, so he couldn't see where my head was. I kept my gun out and walked up and grabbed him by his hair and pulled him out the window, all in one motion. Not knowing if he had another weapon, I dragged him out into the middle of the highway. I took my gun, stuck it right on him, and told him not to move. I got him cuffed and stuffed, then searched him.

My blood pressure was high, but I controlled the situation. It turned out he was going to try and kill me. He had nothing to lose, since he was a Texas Department of Corrections client on parole, as I later found out. I told him, "Today is not your day. I'm just starting my career, and you're not going to f*** it up."

That guy could have ended my career and maybe even my life, but, thanks to Clint, my people skills, my training, the Police Academy, and my foundational skills, I survived. Other officers gave me bits and pieces, but Clint spent a couple of years and provided me with my foundation, so I was able to control the situation. Clint gave me survival training and taught me how to use all my senses. Every step of the way during this stop that training kept me in control of the situation. Had I handled that situation differently, I might now be dead or seriously maimed. I had learned not to put myself in a position where I couldn't take care of myself or other people. I didn't simply walk up to a car and start chatting with someone who had every intention of killing me. Instead, I had noticed something was wrong, got the man's attention, diverted that attention from where I really was, and finally got control of his hands and cuffed him. I neutralized a dangerous criminal and kept myself safe—a job well done. This was repeated hundreds of times in my policing career!

Study Abroad

While I attended college at Sam Houston State University, the premier criminal justice college in the United States, I studied abroad in West Germany, Switzerland, France, and a few other places, at their criminal justice institutions. I logged nine credit hours. The experience opened my eyes to doing things differently, and I was able to see how to improve what I thought were my best practices. These experiences first introduced me to penal reform and crime prevention.

I visited courtrooms in West Germany and rode patrol with the Polizei there, went to Lucerne, Switzerland, spoke with

probation and parole officers, and visited a terrorist prison—
I did it all. My time abroad taught me that there are ways we in
the United States can do things more effectively and efficiently,
especially when it comes to crime prevention. That time overseas
planted the seeds for my many crime prevention activities that
we'll discuss later in this book.

You Don't Need to Be Here

A year after college, I went back to work at the Clute Police
Department, where I had first worked as a police Explorer, because
I felt like I should give back to the community. All of my study
abroad, my coursework at Sam Houston University, and my work
as a cop up to this point had taught me much, and that prompted
me to share what I had learned. The chief in Clute at the time
was female, which was almost unheard of in those days. She was
a great police chief who appreciated me and that I had a degree,
complemented by studying abroad. In fact, the women I've worked
for have all been great.

One day she told me, "Chuck, you don't need to be here; you
need to go off and do great things." Before she even said that,
I had already been thinking about moving to Colorado—where
I spent many summers growing up—to further my police career
because Colorado was very progressive, both philosophically and
technically. Her simple statement, her urging for me to go and
spread my wings, proved to be a pivotal moment in my life and
career. In that one moment, she gave me permission to go off to
even bigger and better things. Of course, I didn't need her affirmation to go onward and upward, but her validation was nice to have.
I started the relocation process, and a great adventure began!

Yes, S.I.R. and Building My Brand

With Colorado's commitment to innovation, I felt confident I could do something more with my life there. I immediately found work in Lafayette, Colorado, with Police Chief Larry Stallcup hiring me over the phone. In those days, handling such business without laying eyes on someone was virtually unheard of. Chief Stallcup sent my chief in Clute a test for me to take and interviewed me over the phone. He liked me and my resume enough that he didn't need to meet in person. My wife and I packed up our belongings, including our dog and cat, and drove north to Colorado!

Once there, we bought a four-wheel drive truck (because of the snow in Colorado) and settled in. I ended up working there until 1998. I needed to learn how law enforcement operated there, because every place polices a little differently. In addition to learning much while on patrol, I buckled down and learned their statutes and traffic laws, among other things. Much of what they did in Colorado was the same, but much was very different. I am always amazed at how differently things are handled state to state.

Becoming Colorado's Crime Prevention Officer

My move into crime prevention had actually started when I was back in Clute, and within two years of arriving in Lafayette, an opportunity opened up for me to become the crime prevention director for their police department. This experience foreshadowed my eventual move into security. That time in Clute, plus my time overseas, as a Police Explorer, and in college had all propelled me in this new direction as a crime prevention officer. Community-based policing, as taught to me by one of my professors, Dr. Marilyn Moore at Sam Houston State University, just made sense to me.

Because of my journey up to this point, I came to realize how important community policing and community outreach is. We law enforcement officers in the United States are good at reacting to crime—we have the best SWAT teams and investigators, and we always will—but our first mission should really be crime prevention and community outreach.

One of the first programs I put together as crime prevention officer in Lafayette was Halloween Safe Night, which usually took two months to organize. This safe Halloween event that parents could bring their kids to ran each year for years and drew thousands of participants from all over the Denver Metropolitan area. To keep it affordable, we charged 25 cents per ticket to this event, which included 100,000 square feet of haunted houses. Halloween Safe Night provided a safe venue, and we brought in vendors, a radio show, and sponsors. We made it a big event that kids would remember, with the added goal of building positive relationships in their minds regarding law enforcement.

At this time, I also became the Police Explorer advisor, which allowed me to give back to the community. Many of those kids I led as the Explorer advisor are now police commanders in major law enforcement agencies. Even though I was a crime prevention officer, I still went on patrol. And when I was on patrol, people would recognize me, which fostered their cooperation. Because I knew them and they knew and liked me, I knew better what to say and how to act to get them to talk and reveal information.

Also, since I was familiar to them, it was easier for criminals to confess to me. This familiarity also made it less likely that I would get beat up or shot. As well, I visited dozens of schools where I met and talked to thousands of kids. Because of this, whenever I rolled onto a scene, I got more cooperation from the people there because they already knew and trusted me.

When I saw the success my rapport gave me with the people I met on patrol, I realized the power of citizens and cops knowing one another. Because of this, I pushed forward a plan to re-define our districts so officers would be permanently assigned to one district, which would allow the people who lived there to get to know the cop and vice versa. While I never got to do that because of police politics and their habit of being reactive rather than proactive, it was and still is a good idea.

The Move to Security

While I was a crime prevention officer, I visited every single business owner, one at a time. I wanted to get to know the people in my community and engage in effective community policing. These visits also made me popular. During them, I took the time to shake people's hands, give them my business card, and ask them what issues they were dealing with.

As I visited these businesses, the owners kept talking about their security concerns. Most of these business owners who had these concerns were in retail, a few in big box chains especially, and they were most concerned with loss prevention, meaning that any security measures they employed were always aimed at preventing theft.

Thanks to my successful work in community policing, I soon got awards, set industry trends, and became the number one guy in the crime prevention officer space. And part of this happened because I networked and became interested in businesses and their security and risk concerns.

While I visited these businesses, I noticed other issues as well, including risks like armed robberies and other kinds of violent crime. I realized if I was going to help businesses with their security concerns, I would need to understand risk from their perspective. With this in mind, I began making moves to become a security professional, and, as part of that endeavor, I joined ASIS International. As a result of joining them and looking at business from a business owner's viewpoint, I started to understand security from a business perspective, rather than just a law enforcement one.

During these visits, I came to know a particular pawn shop owner, Todd, who ran his business with his wife—whom I am still friends with to this day, almost forty years later. It was 1986, and Todd faced huge problems with internal and external theft. He worked full time for a publicly-traded pawn company and also worked on the side for his own pawn business. Todd figured that, because I was in law enforcement, I could help him, so he reached out. I was interested and wanted to help, but I also felt like I needed to know more about the security industry in order to better assist him.

YES, S.I.R. AND BUILDING MY BRAND

To that end, I decided to become an ASIS-certified professional, earning my accreditation as a Certified Protection Professional (CPP) from them, which is the gold standard for security professionals. Getting that certification meant learning a large amount of industry-specific knowledge about security. Also, getting that certification was unheard for a law enforcement officer to do—I had to fly to Dallas, Texas, on my own dime to take the test and get certified. I got permission to do security work as an off-duty job and started working for that publicly-traded pawn shop company under an assumed name as their ad hoc security director. Working in that role, I learned even more about loss prevention, internal theft, insurance, workers' compensation, how to run a business, profit and loss, EBITA, the security measures that pawn shops had to take, you name it. I learned a lot, and it marked the beginning of my move into private security. I could now talk competently to other businesses in town about business and security.

Working with Businesses on Crime Prevention

As part of my crime prevention initiatives, I solicited the logistical support of many business owners, in particular, Randy Strimple, who owned a bunch of Sonic Drive-Ins. He donated thousands of hamburgers to feed our Explorers. What law enforcement calls crime, the security community—private industry—calls risk. And we helped businesses like his see that supporting us would lower their risk.

My focus was to take law enforcement into community-based policing, which had been a big deal at the time. We had everything: D.A.R.E., Neighborhood Watch, McGruff the Crime Dog, Nancy Reagan's Just Say No program, even a police department donkey basketball tournament to raise money. We also had a program called C.O.P.S., which stood for Community Oriented Policing Services.

Everything we did involved the community. We had senior citizens making blankets, toys, and stuffed animals. We'd then vacuum-seal those blankets and toys and put them in police cars for officers to give to people at the scene, especially if it was cold outside. That way, people in distress or at crime scenes would get a warm, custom-made blanket they could keep for life. In our town, officers gave almost every kid a McGruff crime dog badge, and, more importantly, they put that badge on the kid to let them know they were special. It was all about changing minds and attitudes.

My focus on crime prevention got me a ton of awards, including citizen of the year and an award from Roy Romer, the governor of Colorado. I was also the top crime prevention officer statewide and worldwide. My career was picking up speed at a big rate, and because of everything I had learned from being a community police officer, especially the benefits I had seen from more effective community/police relations, I took the strategy and intelligence I had acquired and reinvented myself. It was also around this time that I began to transition into security.

My Security Career Takes Off

Knowledgeable of both security and law enforcement, I started going to conferences so I could learn more about the security industry. As you know, conferences fill their main showroom floors with corporate displays. While at these conferences, I often noticed flaws and gaps in the presentations. When I told those vendors what I saw, they would acknowledge they hadn't thought of whatever I had noticed. "And by the way," they'd ask, "Who are you?"

I would then tell them who I was and that I was in law enforcement and security—basically networking with them—and with those introductions, I started making connections, which eventually turned into consulting gigs.

YES, S.I.R. AND BUILDING MY BRAND

Visiting every business in my city and doing that kind of networking taught me the business side of security and how security and risk affects businesses (versus how security and risk affects crime). Understanding every facet of business, combined with my time spent in community policing, led me to the next phase of my career, which focused on security.

Creating the Package

As my career in security began to grow and evolve, I was often presented with new and unexpected opportunities. At the same time, having become a Certified Protection Professional (CPP), which, as I have said, is the gold standard for security professionals, helped move me forward. I maintained what I had already achieved and continued to deliver a consistent level of service to my clients, while continuing to network and build relationships.

As my career advanced, I built relationships all over the world. And that's why people want to do business with someone, because of the relationships built and the good reputation someone has. While consulting with businesses in security, I learned to be more effective as a police officer because I understood businesses and their needs. I learned about business issues such as profit and loss, human resources, and loss prevention, and this knowledge groomed me to eventually become the chief security officer for billion-dollar companies, all as I worked to become the world's number one security influencer.

Even though I wasn't aware of it at the time, by this point in my career, I was already branding myself. And because I had been a popular crime prevention officer, I was better able to do my job whenever I went on patrol. I learned the power that comes from being known in a positive light.

COPS

Working as a crime prevention officer opened many doors for me. One of the most unique experiences happened when I worked with *COPS*, the popular reality TV show. My involvement with *COPS* all started back when I developed a board game called "Cops the Game," with one of my friends from Los Angeles. (It's important to note that our board game had nothing to do with *COPS*, the TV show.) We sold about 10,000 copies of the game, and it was one of my business ventures.

Knowing full well that *COPS* (the TV show) was the number one reality show in the country, I took the game and contacted Barbour-Langley Productions—the studio that produced *COPS*. After they saw it, they invited me to Los Angeles, so I hopped on a plane, hoping to be able to tie my game in with their show. When I got there, the fact that I had developed the game (plus my experience as a patrol officer and security professional, perhaps) prompted them to hire me on the spot as a technical advisor to the show. You could also say I got that gig because of the relationships I had and because the people involved in the show thought enough of me to offer me the job.

I grew up in the school that says, "If you don't ask, you don't get." Guaranteed. When I was in Los Angeles, I asked the head of the studio, "What's it going to take to make my game *the* official game of the greatest reality TV show in the history of television?"

He replied that my idea was pretty cool, and it would cost me $250,000 for the rights. Yes, a mere $250,000. Suffice it to say, he was eventually able to waive that fee and asked, instead, for a royalty of our revenue going forward if we wanted the marketing tie-in.

With the blessing of the producers of *COPS*, I then went to Walmart, telling them that they could sell a game officially endorsed by the show *COPS*, but they passed, due to their shelf-space calculations. Basically, they made a gut decision because they didn't think my game would sell well enough to justify the shelf space it would consume. Are you kidding me? *COPS* was *the* top-rated reality show with millions of viewers! But that's how it went down. That opportunity to sell board games through Walmart was so good that, if it had worked out, I probably would have stayed in game production and never fully segued into security. It was a golden ticket opportunity that turned to foil—like many do.

But where one door closed, another opened. I was still working at *COPS* as a law enforcement advisor, which helped me grow my reputation and advance my brand. All of this happened when I was in Lafayette, and I perfected my ability to multi-task and run all of these events and projects in parallel. Wanting to keep my hand in police work, during this time I also worked for ten years off-duty for the University of Colorado Boulder (aka the Buffaloes) police department.

In 1998, I left Colorado and returned to Texas to work as police chief for a small community that needed some help. I turned their department around and handed it back to the sheriff's office for service. As my interest in the security world grew, I realized I couldn't achieve my career goals in that sphere unless I truly understood business on a global scale. I then decided to leave law enforcement and go work for Dow Chemical for the next five years (while I kept my license as a Texas peace officer as a Reserve member) to hone my business acumen. While at Dow, I kept my law enforcement and security connections alive by being involved in various security-related projects, but I really needed to understand business methodology. I had come to learn you can't be the chief security officer (CSO) for a transnational, billion-dollar corporation without knowing EBITA, profit and loss, supply chain, human resources, change management, and all that stuff.

Working for Dow and honing my business acumen was a strategic move on my part because Dow was one of the best-run companies in the world. They even sent me to school to get a degree in chemical manufacturing, which they paid for.

During my fourth or fifth year at Dow, that pawn shop owner in Lafayette, Todd, owned and operated a chain of multimillion-dollar pawn shop operations. One day I got a call from him.

"Chuck! What's it going to take to get you back to be my senior VP and chief security officer?"

"I'm already packing my car!" I replied.

Todd quadrupled my salary and paid for everything. I worked eighty hours a week for two years, helping him build that business. Then we sold it. If that's not the benefit of a relationship, then I don't know what is.

First Data

After we sold Todd's pawn company, I wondered what my next move would be. First, Houston-based First Data-Telecheck heard I had left my position as Todd's CSO, so they hired me as security director. Back to Houston I went. Again, the friendships I forged and the relationships I developed along the way helped make this happen.

Around this time, social media—MySpace and LinkedIn—started, and I realized these platforms were where the real power of relationships lay. I was on LinkedIn from day one, which is the greatest thing since sliced bread, and I started building a quality database of people around the world, thanks in part to all the connections I made while still at First Data-Telecheck. While there, I had been at the top of my game, directing security for eight million merchants and banks.

YES, S.I.R. AND BUILDING MY BRAND

I soon found myself sitting on advisory boards at lots of companies, doing all kinds of security advising for a variety of large companies. Because I was also in the financial fraud space, I became a Certified Financial Examiner (CFE), and I got certified as a Certified Financial Crimes Investigator (CFCI). I became involved again with the Houston ASIS chapter, eventually working my way up to their international board within just six or seven years.

In a way, First Data-Telecheck allowed me to continue and expand my life's work, which started when I was a Law Enforcement Police Explorer. Because of working at First Data, others now knew me as one of the top people in the industry, and this allowed me to take my career to another level.

Everything I had experienced so far as a police officer and security professional had prepared me for my next phase. From my time spent as a Police Explorer, to my time as a cop, police chief, and commander, as well as my work for Todd's company, I was building experience and developing relationships. All this, plus working for First Data-Telecheck and gaining experience in the corporate world, along with becoming active on LinkedIn, helped me establish my very own networking organization—Friends of Chuck (FOC)!

YES, S.I.R.

A whopping $4.88/hr as a part-time dispatcher in high school.

YES, S.I.R. AND BUILDING MY BRAND

Officer Clint Hackney putting us through training paces of weapon nomenclature.

Chris, Chris, Chuck (3 Cs): Preparing for weapon training.

The Clute Police Department's First Police Explorer Vehicle, 1976.

YES, S.I.R. AND BUILDING MY BRAND

Facts photos by TAYLOR JOHNSON

PATROL CAR PASSENGER — Chuck Andrews demonstrates his favorite activity with Law Enforcement Explorer Post 396 at the Lake Jackson Police Department. Andrews enjoys accompanying officers while they make their rounds.

YES, S.I.R.

Going on Patrol 1977 with the Clute Police Department Explorer Post 351.

First Day of Patrol Richwood PD, Shiny Badge
and All, Working for Chief Glen Patton.

CHAPTER 5

Friends of Chuck (FOC)

Friends of Chuck is my networking organization, which helps security industry professionals connect with new people, deepen their existing connections, and build better working relationships, all while they help each other reach their goals. FOC also helps security industry professionals exchange intelligence about job openings, industry trends, and whatever else they need. Personally, FOC is the result of my many years of dedicated networking and helping others within this industry to dream big, have fun, and get stuff done.

FOC all started during a weekly meeting of people in the security business—leaders, colleagues, and friends—who would get together in Houston to talk and share ideas. We gathered as colleagues to solve the world's problems, so to speak. During our meetings, someone would invariably mention in their travels how they had recently spoken to someone in Dubai at two o'clock in the morning and how that person in Dubai said they knew me.

YES, S.I.R.

In fact, everyone in the group agreed that everyone knew me and that I seemed to know everyone. It was the same when I was security director for First Data-Telecheck in Houston, because people there would constantly rib me.

For instance, someone in my security circles would travel, often overseas, and as soon as they'd tell someone they were from Houston or Texas, the person they were talking to would ask, "So, do you know that guy Chuck?" Again, my habit of constantly connecting and staying connected, both via social media and otherwise, had laid this foundation.

One of our group members was a guy named Mike, who was a couple years older than me and a former United States Marine and cop in Texas and New York. Mike had always wanted to start a creative promotions company—hats, tee-shirts, promotional items, etc.—which he later named Storyteller Promotions.

One day Mike mentioned that, no matter where he went, he'd meet someone who knew me, and he thought I should do something with that. Then, my friend Frank piped up and said, "Hey, what about calling it 'Friends of Chuck'?" We all liked the idea, and that's kind of how it started. My new venture needed a logo, so Mike said, "Look, let me have my son play with this." And Mike got his son, Matthew, to design the logo for Friends of Chuck in about forty-eight hours, the logo we still use today.

My next task was to decide what to do with this new thing, Friends of Chuck. I didn't feel it was appropriate to charge for it, and I knew I wanted something simple, so I built all these pieces slowly, adding as I went. I learned everything I could about affinity partnerships, branding, and marketing collateral—the whole nine yards. I added promotional items, a website, and an app, and I started hosting events and attending conferences while wearing my FOC cowboy hat, complete with logo. I also added a Friends of Chuck pin and sticker. The promotional items I created allow

FRIENDS OF CHUCK (FOC)

friends to "carry" me around physically, and the app allows friends to carry me around digitally in a pocket or purse!

To take full advantage of social media and its outreach, I started hash-tagging FOC in all of my communications, and then I added an FOC group on LinkedIn. I started hosting FOC events, like social hours and eventually Texas Night, as well as other events. Incidentally, Texas Night turned out to be the biggest pre-opening network event in the history of ASIS International GSX over its sixty-two year history...that we are aware of! Soon, people started to realize that my own database had this incredible connectivity and credibility, and that's how I designed S.I.R.—Strategy, Intelligence, and Relationships. Because strategic, intelligent relationships combine to help people succeed, get stuff done, and develop products—it's a business philosophy. Good strategies must be carried out based on intelligence (intel for short) and fed by relationships. In turn, relationships feed intel, which feeds strategy—all three work together.

Our Mantra

Being part of FOC means that you know Chuck, Chuck knows you, and you're part of a huge network. It's all about friendships because we all believe in dreaming big, having fun, and getting stuff done. That's our mantra!

Now, why did I choose that mantra? Well, nobody can argue against any of those three elements; they're very common sense and essential. First, why dream big? Because who wants to dream small? If you dream small, you get small, right? I'm from Texas, so we do everything big. Second, why have fun? Well, if what you're doing isn't fun, then why are you doing it? Why do something if you hate it? And lastly, get stuff done. We all need to get stuff done because getting stuff done helps you feel valued; why be a part of something that doesn't accomplish anything? No matter what pursuit you're in, you want to get stuff done and accomplish things. It's part of why we're here. Again, these three mantras are hard to argue against and cover all the main reasons we need to get going and move forward.

Relationships and Friendships are Key

Simply put, we call it Friends of Chuck because friendships are relationships, and relationships are friendships. Everybody likes Chuck, so everybody wants to be associated with me. Whether people in the security arena need help landing a job, being mentored, closing a business deal, or discovering emerging technologies, FOC can help them.

It doesn't matter what business you're in; no amount of money can replace or dismiss human interaction and relationships. People buy from people. Yes, your product has to be good, but without relationships you won't succeed. Friends of Chuck is the result of my attempt to build formality around relationships and networking, and it's taken only six or seven years to grow from a few friends to over 100,000 members in over a hundred countries. With FOC, I've created a platform and ecosystem by which I can connect with others and through which they can connect with each other within the security industry. FOC is an institution now within the security world with a building legacy.

When you network, you start to build formality around your relationships, without having to charge for it, and those relationships then become your currency. The network I built up over the years has helped my career, especially when I worked for First Data. Back then, if I needed something, I could pick up the phone and make it happen. Since then, Friends of Chuck has made that even easier. Today, people who are part of FOC just assume they're

FRIENDS OF CHUCK (FOC)

friends of mine, so FOC has taken on its own dynamic. But FOC is not all about me. In fact, I really want others to look at what I've done and do the same, to build their own friendship network.

That's what the rest of this book is about—building your own career and network while helping yourself and others to achieve their big dreams. And the key driving methodology behind that is my methodology of strategy, intelligence, and relationships or S.I.R.

Chapter 6

Yes, S.I.R.

In this chapter, I want to introduce the main concept that ties this entire book together: Strategy, Intelligence, and Relationships, or S.I.R. is a simple, actionable methodology that all security and law enforcement professionals, and, for that matter, anyone doing anything in business, can follow to find success and fulfillment in work and in life. If you live and work by these three principles as I have done, you will be astonished at what you can achieve as you dream big, have fun, and get stuff done.

Strategy

Your strategy is your overall plan. It's how you move forward, and it provides the framework by which you use the other two elements of this methodology—intelligence and relationships. Without a strategy, you are like a ship without a map and compass, a plane without navigational instruments. Your strategy directs how you will best utilize your intel and your relationships. Specifically, it covers everything from who you talk to at conferences, how you engage on social media, to how you build credibility online (and in person) to become an influencer in the security world. You must have a good strategy—that is subject to being refined as needed—to fully utilize the other pillars of this methodology. Want to dream big, have fun, and get stuff done? Great. But you need a strategy to do those three things.

Intelligence

Simply put, intelligence is accurate, actionable information. Intel is the fuel that enables you to execute your strategy by showing you which relationships to build and which actions to take to achieve your goals. Each of us has only 168 hours in a week, and good intel helps you to use each of your waking hours to their utmost. Without good intel, you don't know who to trust or turn to, what's coming down the road, or where the best opportunities are. Intel helps you as you plan and execute your strategy and foster the right relationships.

Intel also means that the information you pass on to others—those people you are in key relationships with—is accurate, actionable, and helpful. You should always be gathering intel, both in your daily reading and social media habits as well as in your relationships and networking. If you give others bad intel, they will eventually find out, and people will stop trusting and engaging with you. You need good intel to execute your overall strategy, and you need it to get stuff done as you dream bigger dreams.

Relationships

While all three elements of this methodology are important, relationships are the most important because, without them, no amount of big dreams, well-planned strategy, hard work, or intel will get you where you want to go. You need other people in order to succeed, and that means you need to develop good relationships with trusted people in the security world.

Cultivating and deepening relationships doesn't only mean you learn who to turn to, it also means becoming someone others can turn to. Relationships are as much about giving—maybe more so—than about getting what you want.

When you cultivate and maintain your business relationships, you open yourself up to help and to be helped. Whether it's finding a job—or helping someone else find a job—passing on actionable intel, or helping the firm you work for get stuff done and doing your job, your relationships are your key to success and to dreaming big, having fun, and getting stuff done. Relationships are king.

Now that you've read and digested this brief introduction, it's not too soon to start seeing how and where these three elements—strategy, intelligence, and relationships—work together to help achieve your goals.

Dream Big

In Texas, everything is big: our economy, our barbecues, and especially our dreams. If you want to get anywhere in the security space, you must dream big, and you won't achieve anything big without making your dreams big. If you dream small, you'll be small. Dreaming big isn't only a nice idea, it's a tactical move that will put you head and shoulders above your peers and help take you places in your career you can now only dream about.

Think of it this way: we're all a bit like fish. In order to grow, we need to leave our fishbowl and swim in an ocean.

In other words, if you want to succeed in life, you need to dream big. And just like a fish needs to live in the ocean to grow bigger, you won't achieve more than you are now until you dream bigger. Now, simply because you dream big doesn't mean you will hit all your big dreams out of the park, but if you never dream big, you can be sure you will never win big.

So, why dream big? Well, to achieve whatever you want, you always have to start at 125 percent, to factor in waste and mistakes, or cannon fodder as I call it. Because your dream is going to have to go through filters and barriers (like time, travel, and expenses), you have to aim high, plus 25 percent, and that equals a big dream. And when the dust settles, you end up achieving in the 90 to 100 percent range.

Let's say I'm a rank-and-file security officer, and one day I dream of being the chief security officer of a transnational, multibillion dollar corporation with assets and operations across the globe. So, how do you land such a sweet, awesome job? Well, to go big you need to dream big, and you have to envision yourself there and hang out with people who work there so you can understand the culture, especially the words they use.

What Should Your Dream Be?

Your dream is always yours and yours alone. It should capture your priorities and incorporate everything important to you. Before you dream big, you don't realize that there's an ocean filled with whales. But when you dream big, you leave your little pond and venture out into the ocean, and it's magnificent. You'll see whales that are 100-feet long, while you're still just an 18-inch mackerel. You'll meet people who are in charge of huge corporations and who earn half a million per year plus bonuses. And here's a phrase to live by: You don't know what you could be or could have until you get out to a security networking event and see who's there and the opportunities that exist. You'll never know any of that until you dive in. Listen, nobody is going to knock on your door and offer you a $300K job at Microsoft. So, if you want to win, you have to show up.

Once you have your big dream, that dream will help direct all your necessary actions. First, your big dream helps you exude the confidence that you're capable of fulfilling those big dreams. Then, your big dream helps guide you to take the necessary steps and make the necessary sacrifices to get what you want. Instead of putting off getting certifications, you're going to get them, and get them sooner than others. Additionally, as you follow your big dream, you will see the wisdom in understanding all the elements of business, like supply chain and profit-and-loss, and knowing those subjects will help prepare you for your dream position. Every step of the way, your big dream will guide you, propelling you forward to take whatever steps are necessary to help you capture it. If you don't stretch yourself to swim in a larger body of water, you won't become the big fish you were meant to be.

Again, start with your big dream, write it down, keep it top of mind, and then let that dream inspire and motivate you to take the necessary actions and make the necessary sacrifices to bring that dream to life. You have to think big if you want to be big. It's just that simple.

How to Dream Big

Now that you have your big dream, let's align your actions with your dream. To do that, here are three tasks or areas you need to tackle:

- Networking
- Building relationships
- Creating/finding mentors

Regarding networking, you need to start associating and networking with like people within the industry, people who can help fulfill your dreams. This means you network with associations like Friends of Chuck (FOC) and groups like ASIS International and others who are already where you want to be. Then, build relationships with these people so you can achieve the third item, which is forging mentor relationships with people who can guide you on how to fulfill your big dreams. And, finally, the mentors you find will tell you all you need to know and do to become as they are.

Dreaming Big in Action

One of my employees at First Data was in a job that really wasn't the best fit for him. He was doing fine, and there were no issues, but he needed to go off and do bigger things and dream bigger dreams.

To help him dream big, I pulled him into my office one day, just as my police chief in Texas had done for me many years before, and said, "Hey, Joe. You don't need to be here anymore, so I'm going to help you become the security director of one of the biggest shopping malls in the United States."

He was a former Marine and loved to get his hands dirty, be in the field, run operations, that kind of thing, and this job was very hands-on and a great fit. He said okay, and within a week he had gotten the job. Now he is one of the most successful and longest-tenured security directors of class A malls in the United States. In essence, I helped him dream big by getting that job as security director, a job he might not have thought he could get.

I helped another guy dream big, and along the way helped him land three jobs, with a little help from the FOC network. I coached him right after he had gotten out of the military. He had huge dreams of being a security manager but only had military Security Police experience. I helped him dream big and plugged him into the FOC network. We landed him interviews with people of influence, which led to three positions, one of which was working as a security manager at Caterpillar Corporation. Yes, I helped connect him, but first he had to dream big.

Whenever I've helped someone find a job, I usually follow the same steps. First, I tell people to find what they qualify 90 percent for and then contact me. If I don't know the CSO or hiring manager, then I know someone who does. I also work with them on their resume. The whole object is to help get them an interview by personally speaking to the candidate's personality. The power of networking goes hand-in-hand with dreaming big.

For instance, when someone you meet at a networking event is helping you dream big, they're doing so because they know you and your personality. So, when they call that CSO who is hiring for a job you want, they can say something about how long they've known you and how great your personality is, etc. These are the pieces you can't always talk about with the human resources department but that are important ingredients for making a hiring decision. When I help someone, I'm usually able to speak to their emotional intelligence as well, which I was able to do for these two people I helped.

What to Do

To dream big, here are three actions you need to take:

1) Join ASIS International and become a Certified Protection Professional, or CPP, which is the gold standard in the industry. Your mentors should be able to tell you how to obtain that certification as well as how to use it to further your goals. Join other security associations, too, based on the future role you aspire to—Security Industry Association (SIA), Association of Certified Fraud Examiners (ACFE), etc.
2) Find a mentor who already has the life and career you want, and let them guide you on everything from which

conferences to attend down to which hotel to stay in (which is always the host hotel where the big dogs all stay).
3) Network and hustle. Get your name out there, building your reputation one relationship at a time. This is about "quality" not "quantity." You should do this by attending live networking events and by posting on social media. Your goal here is to make yourself relevant to the industry. When you post, don't just make noise, but post about topics or news items that are important and relevant to the industry. You also need to respond to others' comments and generally contribute. If you want to be a CSO someday, then start acting like a CSO.

The Power of Live Networking

I believe there is no success by Zoom—you have to show up. If you want to dream big and succeed, there is no substitute for live networking events. Once there, you want to engage with people who live the life you want and work in the kind of job you want. Then you need to emulate them so you can see and think as they do in order to become like them.

Everything you do at that event matters, from where you wear your pin to how you shake hands to how you converse with others. Watch how the successful people dress and act and do likewise. Networking allows you to surround yourself with other like-minded, successful people, and soon you'll start to be successful like them because success is contagious.

Recapping the Steps to Dream Big

To help drill these steps and process home, I want to recap the basic steps to dreaming big:

Step one: Show up and be around people who are bigger and think bigger. Be challenged by others whose successful attitudes are contagious. And by showing up I mean physically attend security network events and work the room, so you can meet the right influencers and then connect with them—you're not just there for the free chicken dinner. P.S.: *Do not* forget the business card!

Step two: Find an excellent mentor from among the big people you now know. Finding the right mentor is the difference between lightning and the lightning bug. When you pick the right person, they will help you create the right strategy, gather the best intel, and connect you to the right person (i.e., relationships). They will inform your thinking and help guide you to achieve your big dream. They're going to open up doors, thoughts, and opportunities you never dreamed of.

Step three: Make sure your big dream fits your overall values and aspirations. Schedule a weekly discussion with your mentor or coach, but hold yourself accountable to your big dream. Write your dream down, put it on video, keep a video log, or write your dream (or dreams) on Post-its® and stick them to the bathroom mirror you see every morning. However you document it, the important thing is to put it down on paper to better hold yourself accountable. If you want to swim with the whales, then you have to start by hanging out with them. Holding yourself accountable—along with all of the other advice in this book—may seem like a sacrifice, but one of my personal quotes is, "Life's not so much about choices as it is about sacrifices."

Again, relationships are key, especially when it comes to dreaming big. Relationships evolve into mentors who can then help you dream big and fulfill those big dreams. Your mentor relationship can be formal or informal, but it always starts with a friendship. Dream big, have fun, and get stuff done. And all of these happen better through relationships. Next up? Having fun. Because who doesn't want to have fun? See you in the next "fun" chapter.

Chapter 8

Have Fun

We in law enforcement and security deal daily with the devil's business—we see the worst humanity can offer, and we see it again and again. It's not surprising, then, that many of us have trouble having fun. But the truth is, this industry *can* be fun. If you want to succeed in it, you have to have fun wherever you are and whatever work you're doing.

We all know life is hard, but having fun helps the day go quicker and helps us live happier, more successful lives. We are also more productive! Thus, one of the key secrets to being successful in the security space is to have fun.

Know How to Have Fun

Law enforcement and security work are serious businesses, and we cops can get pretty morbid: we become cynical, start to dislike people, and too many of us turn to alcohol and/or drugs to dull the pain, only to end up turning over our lives and freedom to addictive or bad behaviors. We in law enforcement and security are also in the problem-solving business, and that can be negative too, making having fun a challenge. Especially in the security business, you're dealing with risk, urgency, critical risk, being responsible for assets and supply chains, etc., and everyone looks to you for answers and protection. If you don't learn how to have fun, you won't last long. Instead, you'll burn out, lose out on career advancements, and fail to achieve your big dreams and success. Forget how to have fun and everyone you meet will be able to tell it from miles away. So, it's essential you learn how to have fun while doing your job and getting stuff done.

Believe it or not, not everyone knows how to have fun. Maybe you grew up in a horrid environment and/or had to work to survive from a young age, so having fun wasn't something you had a chance to learn. If you don't know how to have fun, you might not be to blame. But you still need to learn what fun is and how to have it if you want to succeed in the security industry.

Why is it important to learn how to have fun? If you don't, you won't achieve your big dream—remember it? And part of your big dream should be to have fun at what you do. Just as a smile is contagious, so is having fun. When others notice you having fun at your work, they start to have fun, too. It spreads and people want to be around you, which can lead to more advancement and more success. Again, having fun equals success.

Choosing a Fun Mindset

Now that you know how important having fun is and that it is a practical, tactical method to achieving your goals of being a leader in the security industry—and to maintaining your mental health—I want to talk about its key ingredient: choosing a fun mindset. Being able to have fun in this business all hinges on having the mindset of making your work fun.

An essential element to creating a fun mindset is to determine that, no matter what you do, you can have fun doing it. This isn't a far-off wish but a conscious decision. One technique to help create this mindset can come from testing new ideas or job descriptions, especially when you're considering a change or are offered a new position. For instance, if you're entertaining a job that requires you to be in a closed office in a basement downtown in a major metropolitan area answering emails all day, clearly, this is *not* fun if you're a hands-on in-the-field type person! In other words, each time you consider something new, look at the totality of the opportunity and ensure it allows for "fun," which means, you must be true to you. If it doesn't feel or look like it might be fun, it likely won't be. And if you don't have fun, your big dreams will fade away into little dreams—or no dreams—and you won't be successful.

Having fun also helps you be successful because it is contagious, and your fun attitude will spread to everyone else. After all—and this may sound obvious—everybody wants to have fun. When you approach a new assignment or position, make the habit of looking for the fun in it. Consider the relationships you are going to make, the credibility you're going to build, and the success you're going to have because of this new assignment. Nobody wants to be around a negative Nelly or a naysayer, so do what you can, wherever you are, to find the fun in what you do. Your success depends on it.

How to Have Fun—An Overview

Having fun isn't just for kids; it's essential. In this line of work, being able to have fun can enable you to have a longer, happier life and a more fulfilling career. Having fun will also help you be more successful, as it keeps you fresh and alive. Because having fun has been an irreplaceable part of my success, it's essential you learn some basic steps to follow to have fun, especially if you tend toward the negative.

First, figure yourself out. You're not good to anyone if you're not well, mentally. If you can't be who you are, you're living a lie. Being in this arena is a lot like being in a marriage, because you're married to this job, so you have to be your best. When you're not your best, you are no good to anyone, including your wife, kids, or your job. So, make sure to get yourself in the right place if you aren't. If you're in a bad situation, look at yourself and make sure you can be all that you can be. This could even entail looking for another job, if that's what will help you get right.

Second, create some contrast and context. If you are unhappy in your current job, be it the low pay, bad boss, or too much travel, make sure you know where you are before you make a change.

Third, having fun all starts and ends with you, so have conversations with people around you to understand how you can make improvements. This is where organizations like FOC come in because they help you connect with other opportunities that can ultimately further your career and help you have more fun.

Tips to Help You Have Fun

If you feel you're in the right situation but you're still finding it difficult to have fun, try following these tips. For starters, ask yourself *why* you aren't having fun. Name the three things causing you not to have fun. If you can do something about them, then do it. You can also discover what things you can do or change that will make your job more fun. Write these items down because doing so will help focus your thinking.

It also helps to find some perspective. Ask yourself how bad your situation really is. Sometimes, a situation becomes less dire when put into context and contrasted with what else might be happening. You have to create the context and contrast it to appreciate what you have. Because you don't know what you don't know, you need to get out there into the larger ocean to see what else is out there. There may be a much better life waiting for you, but you don't know because you haven't gone exploring in bigger waters yet.

The Practical Value of Fun

I recall a security consulting engagement I had as a young man where I was not having fun, nor could I make it fun. Despite my efforts to boost the company culture in a positive direction, the environment and leadership were toxic.

The money was good, really good, but it was clearly not worth it! I tried to use all my skills and past experiences to change attitudes and bring understanding to the mission, but the client would still yell on the phone, send emails with scathing remarks, and mandate unreasonable requests and expectations.

In situations like this, what can you do?

In this case, I fired the "client" and went to "bigger waters". Talent will leave leaders who can't find the practical value of fun and create a positive work environment where people can enjoy themselves while getting stuff done.

Sometimes in the security business, you have to realize there are people and projects that, no matter what you do or try to manage, their personality is so toxic, there is no amount of "fun" allowed and it is not worth your time. You may get to a point like I did where no matter how good the money is, you have to cut ties and find a client that allows for fun.

Fun can show up and be experienced in many different ways. As a security director, I always found myself working far outside the box to ensure my employees had a fun working environment by acting, from time to time, spontaneously.

For example, in one case I gathered my team and off we went to the movies! I remember taking my team to see *Paul Blart: Mall Cop*, complete with popcorn and Coke. We thought of it as a humorous training film.

Needless to say, as a leader, I wanted to set a practice that instilled a "know how to have fun" environment, all while teaching each of them to copy/paste into their own careers. Even today, they've never forgotten that "fun" moment working for me.

Having fun is an essential skill for all leaders in the security business. If you can have fun while getting things done, your recruitment and retention of talent will increase, which impacts the bottom line and adds that much more to the fun you can all have in the form of potential pay raises!

Finding Your Own Fun

Keeping things fun in the security and law enforcement world is not always easy. As for myself, I'm okay with making even horrible stuff funny—because crime scenes can be really awful and finding humor in a situation can help maintain your sanity. As a consultant, I don't take on projects or clients that I suspect won't be fun or that I can't figure out how to make fun. So, unless I can do my mantra (dream big, have fun, and get stuff done), I won't take the gig.

If I'm already in a situation that isn't fun, as when I was working as a corporate security director, I find fun in what I'm doing on the side such as the consulting I did.

If you have a great job you need to keep but it isn't fun, and you are allowed to moonlight a little, then take on side work that

HAVE FUN

you find enjoyable. If you're not having fun and find you can't stay where you are, you may need to make a change.

I counsel people in this industry to ask themselves, maybe every six months or so, to describe three events or things they're doing that are fun. If they can't, it may be time to make a change.

Sometimes you aren't immediately aware of the fact that you are not having fun, as it can creep up on you.

One time, I was going through a lull as a corporate security director, not having much fun. Through my relationships, I attended an ASIS International Young Professionals event and met someone tasked with dealing with some serious national fraud issues. They had no idea what to implement to lower their losses and risk. In short, they invited me to come speak to their board as an SME, and that turned into a compensated consulting gig…which I enjoyed doing! This brought about the "fun" I needed to balance my corporate security life with interesting challenges, which all came from relationships. I've repeated this "fun" activity a hundred times in my career. Secondarily, and as a plus, I grew my relationships and expertise in the fraud business!

Again, ask yourself what drives you, what's fun for you. The answer will be different for each security professional. For some, it's public service or saving the community, and for others it might be implementing innovations or working with new technologies. It doesn't matter what you find fun, just make sure you get to do it during your day.

Remember, if you're not enjoying yourself, you won't be successful. Do you really think someone is going to open a new, golden opportunity to you if you aren't fun? Nope. You have to learn to have fun—I can't say that enough.

And, after dreaming big and having fun, you need to get stuff done, which is what I want to talk about next.

Chapter 9

Get Stuff Done

Everyone needs to feel valued, useful. We all also need to feel like our lives have a purpose and that we're not just down here on Earth killing time. One of the best ways to be useful and feel like you have a purpose is the third element in my mantra: get stuff done.

Almost nothing can make you feel better about yourself than getting stuff done. Whether writing a book, completing a long-delayed task, or building something new, getting stuff done makes you feel good and helps you succeed. As for me, I can sometimes work up to sixteen hours if that's what it takes to feel like I've gotten enough done. All of us want to find success, achieve our dreams, and do something with our lives, which means we've got to get stuff done.

Got big dreams? To achieve those, you've got to get stuff done.

Want a better job? Looks like you've got to get stuff done.

People don't want to hire or hang out with people who don't get stuff done, don't follow up on calls, or don't understand maintenance. So, get off your butt and get stuff done. It's just common sense.

The Importance of Maintenance

One of the key ingredients to getting stuff done is maintenance. Yes, plain, boring old maintenance. Maintenance is key to your success, but it's also something most of us suck at! Businesses suck at it. Humans suck at it. Chemical plants suck at it. It's costly, repetitive, and boring. It can be hard work, too. I mean, who wants to be a maintenance man? But it's the most important job in the world. Maintenance is king. You can be presented with the best opportunity in the world, but if you don't maintain it, it's gone. You can have every door in the world opened to you, but if you don't maintain what you've found once inside, you're gone. You can have all the talents and gifts in the world, but if you don't maintain them, you might as well not have them.

Maintenance is not easy, but it's a huge key to success. Maintenance is a key element to change management as well, so Dow Chemical developed a methodology called management of change (MOC) in maintenance. They documented, down to the millisecond, how to maintain key processes and procedures. And it worked.

When you want to get stuff done, focus on maintenance during the beginning, middle, and end. Maintenance is all about getting stuff done. Because when you don't feel good about yourself, it hurts your psyche, and it takes you into a dark place where you can't be as successful as you could be. And you lose out on life and career opportunities. To successfully maintain your career, you need to uncover the systems, processes, and relationships that can all help you to get stuff done.

How to Get Stuff Done

Getting stuff done isn't complicated, but it takes effort. Three items—systems, processes, and relationships—can help you.

Systems. Here, you develop your own methodology and use technology to get stuff done. This might mean using communication devices. For example, setting up a group text feed where you can keep your team in the loop so you can get more done with less; adding the LinkedIn App to your phone to more quickly answer messages and complete posts; ensuring you have cell phone numbers for your relationships in order to timely communicate with them by voice or text. You can go one step further by noting in your phone contacts how people prefer to communicate, whether by email, text, Facebook, or Messenger.

Processes. To be successful and get stuff done, streamline processes to accomplish more with less time, material, and cost. Make sure the processes you use work for you and your team and that they are scalable and easy to use. Having clarity and being consistent with how you get stuff done pays off in the long run, too.

Relationships. If you have to go through bureaucracy to get what you need done, like finding the company CSO to help you work an investigation or assist in resolving a security risk, that's where the power of organizations like Friends of Chuck (FOC) comes in.

When you've built a relationship with me or someone in FOC, we can pick up the phone and call the CSO of whatever major company you need to contact (like Microsoft, Kellogg's,

Boeing, etc.) to find out who you need to talk to in that company to get your stuff done. That's how you do it.

And this last item, relationships, is the most important of the three because it can quicken what you need done—by months! So, don't forget to use relationships to get stuff done. To illustrate this point, let me relate a story.

Using Relationships to Get Stuff Done

FOC has grown a lot. For every person I know, there are a thousand people who know me and my brand. Seven years ago, I was walking back to my hotel room at a security conference. I was tired and it was late, probably ten o'clock, but as I walked, a guy came up to me, asking for help. As a side note, this conference was being held not long after the hurricane had devastated Puerto Rico. Now, the guy was part of FOC but not someone I knew personally. Long story short, he had a multi-million-dollar contract to deliver security services in Puerto Rico, but he and his company were in danger of losing the contract because he lacked a way (either boat or plane) to move the necessary people and supplies to the devastated area. And his company had no contingency plan in place.

"Can you help us?" he asked. "Just please do whatever it takes." I told him I could help and that I was going to go upstairs and get him a C-130 transport plane.

"What?" he said.

"Yes," I said. "And you'll have it in two hours."

After two hours and a couple phone calls, I had his plane. A while later, to my surprise, his company sent me an unsolicited compensation check for helping them. Moral of the story: I called the right people because I knew the right people and had built relationships with them… such is the power of relationships. His company didn't lose their multi-million-dollar contract all because of relationships developed by the guy wearing a big hat walking down the hallway toward his hotel room at ten o'clock at night.

If you want to get stuff done, the best way is via *relationships*. Hands down.

Get Stuff Done

So, how do I become someone who gets stuff done? How do you become the go-to guy in your organization when others need a solution fast? How do you create such tremendous value that everyone wants to know you? By being the one who gets stuff done. Here are a few steps I follow to do just that:

1. Identify and write down why things are *not* getting done—the five barriers or roadblocks to getting stuff done. The reasons things aren't happening could be lack of time or resources or poor communication. In every engagement, set expectations up front with all concerned parties.

    Whenever someone comes to me for help, I try to address all of these common reasons for failure up front and make sure we address them before diving deeper. Clarity, expectations, and deliverables are key!

2. Break down each reason and figure out which systems, processes, or relationships can help resolve the issues. Maybe the answer is as simple as making a phone call or putting into play a simple process like texting everyone on your team instead of emailing them, for instance.

    Basically, you need to figure out which system, process, or relationship you can employ to help resolve your issue with the greatest response. And always build in at least 25 percent to account for losses, waste, and mistakes.

3. Details matter when you're getting stuff done, so determine your deadline and document everything. When I work, I put everything I do on a calendar. At work and during meetings, I take prolific notes and record the date and time as well as who said what, where, why, and how. I also record detailed information about everyone I meet and form relationships with so I can create a profile about them. This profile information covers everyday stuff like their hobbies and likes to their dog's name, their favorite liquor, etc. (I've probably generated over 400,000 documents in my file system over the past ten years, so if something ever happens to me, there's a record of who I've talked to and what I've done.)
4. Talk to others in the industry and ask what's worked for them. This is the power of networking. Oftentimes, others have already been down the same road, and their help or suggestions can save you time, money, and headaches. If you're not getting stuff done, you should be turning to your security network to find out what you're missing or to get new ideas.
5. Last, but not least, once you build your own list of contacts and relationships, do as I have done and index the hell out of it so you can maintain your connections. After you've made enough connections and gathered enough relationships, you won't be able to keep all of that information in your head, so develop your own system and log that information and make it searchable. This also makes it infinitely easier to maintain these relationships and connections.

My networking has been so successful that I now know so many people that when I go to call someone to get something done, I have to think for a minute about who is the right person to call. In these cases, I often refer to my database of connections, which helps me make the right call to the right person.

Want to get stuff done? Just remember these three things: systems, processes, and relationships, especially relationships. In my next chapter, I talk more in-depth about strategy, intelligence, and relationships, or S.I.R. To kick it off, we'll cover strategy first. Let's go.

Chapter 10

Strategy

Strategy is the brains of your operation, which is why it is the first element of S.I.R. Strategy guides all the other elements of your work toward success. Specifically, it guides how you engage in the other two elements: intelligence and relationships.

Your strategy helps you direct what intel to gather and how, and it guides you as you develop relationships. At the same time, your strategy is informed and can even originate from the intel you've gathered, and your relationships influence and inform your overall strategy, making the three elements of S.I.R. interrelated.

Strategy directs your path toward success, and, to succeed, you must have a good strategy. Strategy helps you get whatever you want, and without a strategy, you are lost in the night. Want a good job? You need a strategy. Want to get products placed? You need a strategy. Want to build friendships? Build a good strategy. No matter what you want in this industry—and in life—you have to have a good strategy to get it.

YES, S.I.R.

Oddly enough, most people working in the security industry don't have a strategy. Let me repeat that: most security professionals, an industry known for teaching strategy and preparation, don't have their own strategy of how to succeed. They are like ships adrift on a huge sea (or maybe just a fishbowl!) without rudders. And because they lack a strategy, they don't dream big, have fun, or get stuff done. If you want to separate yourself from the herd, then you must put a thoughtful strategy in place.

You need a strategy on how to get whatever it is you want, be it a job, product promotion, climbing the corporate ladder, whatever. You need to write down what it's going to take to get what you want. This entails knowing who, what, where, when, why, and how on each person you need to talk to and using that to determine how you will approach them. This is a little like tradecraft that people in the intelligence fields have to use.

Building a Good Strategy

So, how do you build a good strategy? Well, you start with information, or intel, especially from people who are well-placed and well-trusted. The intel can vary, depending on the nature and degree of the relationship, going from a casual acquaintance, moving up to colleague and then to a trusted colleague, to one's best friend and/or family. Besides intel, cultivate your relationships, which will help inform your strategy, and this includes building a sound social media strategy.

Your Social Strategy

When you think about it, social media is amazing. Using just your personal computer or smart phone, you can get and stay in touch with people all over the country and all over the world. Using

STRATEGY

social media, you can get your name out there, stay connected with your relationships, spread your influence, and find work. Using social media, you can make yourself look omnipresent, and, thanks to social media, we can all stay connected with almost all the people we know for the first time in history. Social media can give you a huge strategic advantage over those in the industry who ignore it. Yes, there are a lot of distractions on social media, but if you use it correctly, social media can help propel you ahead of the crowd.

To make social media work for you strategically, first make sure your professional profile (which should be on LinkedIn) is up-to-date and relevant. Include enough information about yourself, but not too much. Your profile needs to help those who view it to feel comfortable with you and see you as approachable. Be findable by listing your email, twitter handle, and maybe your phone number on your profile. When people reach out, respond in a timely manner to show that you're someone who gets stuff done. How you present yourself and how you act on social media becomes your reputation and brand. I recommend having a presence on LinkedIn (for your professional side) and Facebook (for your social side). Log on regularly and respond to comments. You also need to post regularly, but make sure your posts contain relevant, worthwhile content. When you're on social media, you build relationships with people even if you aren't aware of it.

To get people to follow you on LinkedIn, you need to adhere to these guidelines:

- **Be consistent with your picture**—don't keep changing it out. Leave it up for a good while, at least a year or so.
- **Make your picture professional but also current**—when people meet you, you should look like your picture. I'm the only guy in a Stetson, which I wear to events and in my

social media picture. As a result, I'm the most identifiable guy in my industry. Go find your Stetson!
- **Make sure you display as much content about who you are and what you do as you are comfortable with.** Again, don't reveal too much, but if you hide yourself, people won't know who you are and what you do. You don't want to miss any opportunities because people didn't know you were qualified for something.
- **Make your profile public and searchable.** There is a setting within LinkedIn that ensures your profile can be seen by as many people as possible. You can throttle back as needed, too.
- **Make *quality* connections versus *quantity* connections.** Don't accept every connection request because many aren't worth the connection (spammers, spies, etc.). Do your homework!

Why does this all matter, especially if you don't like social media? It matters because the better you align with others on social media, the more relevant you'll be, and the more positively others will perceive you. The goal is that when you eventually meet someone you have connected with online, you are more than halfway to really knowing them and building a meaningful and credible relationship. Instead of starting from square one, you've already established trust, which provides credibility before you even meet in person.

Must-have Daily Social Media Habits

To become relevant and stay relevant, practice these social media habits daily. They will also help you get the most mileage out of your social media presence.

STRATEGY

Read through it. You don't have to spend hours, but stay up on the major posts on your feed every day to get the latest intel and strategic information about the industry and others in it. Remember, it's about maintenance!

Comment as appropriate to your industry. When an influencer in your industry posts and you have something relevant to say, comment because it will appear as a post on others' daily feeds. Commenting also reminds the influencer and others that you exist and establishes that you are engaging and willing to offer knowledge around the topic in a post.

Post good material relevant to your industry and the job you'd like to build or share. Do this at least once per week! When you comment or post, you should do so for any or all of the following three reasons:

- **To find work**—LinkedIn is a great way to connect with open positions and share existing jobs out there.
- **To build your brand**—posting regularly reminds people who you are and lets them know what and how you think.
- **To network**—when you engage on social media, you're basically networking, even though it's not in person.

Quality over quantity. When you're building your own network, remember that quality, not quantity, is going to help you win the long game, both with posting and when connecting to others. You should be reviewing every profile you link to. Every week I reject at least 150 LinkedIn connection requests because they're either foreign national spies, fake, or not related to the industry and only want to "harvest," connect, or sell me something (usually Bitcoin, insurance, coaching, or a franchise).

YES, S.I.R.

Creating a Job-seeking Strategy

Your strategy to find work should be all about networking and building relationships. Make sure you show up to relevant events so you can network and get known. While you're at an event, always have your business cards with you so others will have something tangible to remind them of you later. *Yes*, a business card! Your business card matters, too. It should be unique and contain the right content but still have space for people to write notes on the back. If you don't use relationships to land your next job, you're letting the automatic scanning that companies use to filter out incoming resumes (using optical character recognition [OCR]) decide who they will interview and who they will reject. So, don't let OCR decide if you will get that job. Have a real person help you instead.

When I help someone find a job, they need to do or have done the following:

1. **Their resume has to be tight, well-written, and free of typos.** It needs to succinctly explain their experiences and showcase their strengths. Nowadays, human resources and recruitment professionals recommend that resumes should be no longer than a page, and definitely no more than two pages if absolutely necessary. If you ask me, resumes should be "short and sweet," since studies have shown that anything longer won't likely be read.
2. **They have to already have an established relationship with me**, and I need to already know who they are and what they can do.
3. **They must use Indeed (and other search engines) and set up a profile and online resume.** I encourage them to use Indeed or other job search alert features to automatically

search for the kind of job they want by geography, salary, type of job, qualifications, etc.

If a job posting comes up where they are a 90 percent match, I have them send the job posting or job key number to me. If I think the position is a good fit for them, I can make a phone call to the chief security officer (or other influencer) at that company and recommend the jobseeker as being worthy of an interview. And 90 percent of the time, I can get them an interview.

4. **Once I get the candidate the interview, it's now all on them.** They need to get coaching and rehearse before their interview. In the end, who they are comes out in the interview, so it's up to them to show that they are a good fit for the company. Of course, the candidate will get nowhere if they can't get the interview. As I have said already, if you want to win, show up!

Strategy and Relationships

Whether you get that job or not, it all comes down to relationships. And that is why I tell people to start building those key relationships now. Two years from now when you are ready to make a change, you can leverage that relationship when you go job hunting. Again, when you're job-hunting, you have to make sure your LinkedIn profile—your content, including your picture, background, past experiences, etc.—are all relevant to the work you're seeking. And you need to build and maintain industry relationships because those relationships are investments into landing future work. You will land your best jobs through relationships… always and without exception!

YES, S.I.R.

Dream Job Story

One of the members of FOC, a friend of mine, had a specific big-dream job, the kind of dream job you might see in a movie. I can't say too much about the position because it is for a very famous entity. Anyway, the guy never thought this job would come open—it was highly competitive—and if it did come open, he never thought he would land it. But, like any successful person in security, he had a strategy and he followed it. He was persistent: he kept networking—attending all the right events and showing up and greeting people—and he worked hard at his job and kept getting stuff done. Besides being a good guy who knew how to have fun, he learned, studied, prepared, and had obtained the right certification, his CPP. Again, he had an overall strategy, and he kept following it.

Well, one day that very dream opportunity opened up for him—a once in a lifetime event—and because he had laid all the right groundwork, he was ready to roll. Because he had leveraged his network, which happened to be the Friends of Chuck, he avoided the gatekeeper and the OCR and landed an interview. And because he knew how to interview and had all the right qualifications, he landed the job. The point of this story is that, through the power of relationships and strategy, he was able to get that job. He had a strategy, and because of that strategy he landed his dream job.

Strategy is important, but it won't work without good intel, which feeds both your strategy and your relationships. And intelligence is what I want to talk about next.

When I'm at an FOC event or any networking event, it's hard to go five minutes without running into someone who has achieved big results from being in FOC and other networking groups. Whether it was the next job, the dream job, the key contact, or the big solution…so many have happened through networking, showing up, and getting stuff done.

Intelligence

Intel is actionable information we can use to better understand people, places, opportunities, products, and services. Every time we gain information we can act on, that information becomes intelligence, or intel for short.

However, you can't gather intel unless you have trusted conversations with people. And you can't gather intel unless you know what written sources (online or otherwise) to read. Thus, when you know where to find good intel and how to build good relationships, you gain access to good intel. That intel, combined with relationships and strategy, then becomes even more powerful and can help you dream big and get stuff done. Intel helps you succeed.

As important as it is to have good intel, lacking good intel is a show-stopper. Without good, actionable intel, you won't know who to talk to, how to talk to them, or whether the information you're getting from others is valid. Without good intel, you can't assemble a good strategy and won't forge the right relationships. And without good intel, your chance of achieving your big dreams and landing the position of your dreams all but disappears.

You want to succeed, and to do that you need good intel because good intel will enable you to identify the key people and the real story. Want good intel? Then you need to get it from the horse's mouth, and that only happens after you build strong relationships with those in your network. Intelligence is critical because, if you're not using the right words, attending the right event, or speaking to the right person, then you're just spinning your wheels.

Types of Intelligence

Not surprisingly, there are several types of intel, and you need to know how to gather them to help you succeed. Intel can include spoken testimony, inside industry information, knowledge of job openings, news about something upcoming (but not yet announced), and so on.

Intel can include knowing whether the person you're talking to is really a chief security officer or just a mid-level manager who is handling a small piece of the operation, a decision-maker or just a cog. Intel can include information about someone's background, what committees they sit on, who they work for, and what you and they have in common. When you uncover commonalities, like similar work histories ("We're both cops!"), places served in the military ("We both served in the same division in the Gulf War!"), or other common ground, that intel can strengthen your relationship with them and help you obtain even more and better intel. When you gather and discover these commonalities, you build credibility and deepen the relationship, hopefully turning it into a trusted relationship. This is being human!

INTELLIGENCE

Gathering intel is a bit like building an arsenal made up of actionable information. You need to know what's going on, what's happening and being worked on in your industry, and who the people are that you're meeting. If you want to be involved in something up-and-coming, then you need the right intel. And inside information on the latest startup or the latest technology doesn't just show up on your front doorstep. You have to go out and build trusted relationships and then gather the intelligence from those relationships. Again, do you want to win? Then show up!

Chatham House Rules

Part of growing up is learning when to talk and when to shut up. As you gather intelligence, you have to respect what I call Chatham House rules and know when to talk and when not to. When people entrust carefully guarded information with you, you must guard that information as well. Share it freely, and watch your reputation and standing crumble. Having the right information in your arsenal makes you more attractive to the security community because it means you know people, but you must be wise about with whom you share that intel.

Industry Information

The next vital bit of intelligence you need is industry information: what companies are buying which businesses, what investments are being made, what verticals you should be working in or investing in, and so on. Today, the hot industry is cyberspace or artificial intelligence, tomorrow it might be crypto. There are a few places to gather that industry intelligence, including personal relationships, social media, and industry publications.

Business, especially the security business, is all about people, and when you build strong relationships, especially with those in different industries and verticals, you obtain better insights into your own industry. With new information in hand, you can then vet it against those you know and trust. And the more people you know, the better you are able to sift through the noise and establish good baseline intel. Good intel can also appear on social media, but you have to know how to spot it, which means you have to actually be on social media. As I'll mention shortly, you can also gain good industry intel from reviewing industry periodicals.

Basically, if you want to get ahead and fulfill your big dreams, you need to know what's going on in your industry, and relationships, social media, and other sources such as industry periodicals can help you do that.

Quality Relationships Yield Quality Intel

Intelligence is critical, but it can't happen without relationships. Intelligence can help you find the right person through networking, LinkedIn, and social media. Out of that relationship you also get intelligence. Because you're sharing information as you network, you are sharing who you are and what you know. That's why you need *quality* not *quantity* connections on LinkedIn. Go to my LinkedIn page and look at who follows me, and you'll see that the people who viewed my profile also viewed other top people in my field. It's also a good practice to update your LinkedIn experiences every quarter. LinkedIn is intel, and people are going to look at it, so keep your information updated and accurate. You have to be on top of all of this intel if you want to be successful—jobs, branding, startups, opportunities, etc.

INTELLIGENCE

Where to Get the Best Intelligence

Want to know where to get the best intel—the one place that beats all others, where you can hear the inside scoop on industry trends, learn about key job opportunities, and get one step closer to fulfilling your big dream? It's at in-person events. Yes, you gather the best intel in person because people aren't always going to share this stuff on Zoom or social media. In person, you get the nuances and deepen the relationship in ways you simply can't over Zoom. You get your best intel from showing up and talking face-to-face.

Good intel can come from reading and understanding the best sources and can even come from social media and blogs, but in-person is always best. Many in the industry don't post on LinkedIn, especially if the information is sensitive. And even if they do post on social media, the intel they post is often generic—no one wants to give away privileged information to just anyone, anywhere. Read social media to be sure, but when you want the best intel and the inside scoop, face-to-face is your best source.

Intel Multipliers

Now that you know the kinds of intel, there are a few good, potential sources of intel that you should have on your radar. These include:

- Social media (which we've already mentioned), especially LinkedIn, followed by Facebook (Twitter is too noisy, but good for real-time events)
- *ASIS Security Management* magazine. You must be a member to receive it, but it often contains relevant intelligence.
- *International Security Journal* (ISJ)

- Blogs or connections outside the industry
- Some mailing lists. If the mailing list is via email, use an alternate email address so you don't junk up your inbox, and, if you can, parse out executive summaries so you can separate the gems from the rocks.
- Security Industry Association and other security industry blog posts. I recommend receiving their blogs in summary form.

Wrong Intel Can Harm

There's an old Japanese proverb that states, "The reputation of a thousand years may be determined by the conduct of one hour." Your reputation is precious, and, while you may never act inappropriately, if you don't have the right intel, your reputation and brand can suffer when you relay information that isn't true, even if it's something small. Wrong intel can cause you to coach someone in the wrong direction, ignore a vital warning, or miss a key opportunity. Your reputation and ability to get stuff done depends on good intel, and wrong intel can wreck that.

Fall short in one of the three elements of S.I.R., and the other two suffer as well. People don't hang around those who provide bad intel; they just walk away. Wrong intel can even get you blocked on LinkedIn or removed totally. So, keep your S.I.R. together by always getting and giving good intel.

Daily Habits Regarding Gathering Intel

Gathering intel happens when you form and maintain good information-gathering habits. These include continually networking, being respectful, good diplomacy, and always connecting. (And, as I've said above, know and follow the Chatham House rules by knowing who to do business with and who not to do business with.) Part of gathering intel is to read daily from the best sources, both inside and outside the industry. You need to know what has been happening, and it takes time and discipline to do this.

Here then are a few habits you must form in order to always be gathering good intel (if these sound like hard work, they are, but they're worth it):

- Get on LinkedIn daily and review your feed. Read ten posts from ten different connections at minimum.
- Read five to ten security industry periodicals and blogs daily.
- Schedule calls with key LinkedIn connections to build and maintain relationships.
- Read at least one respected newspaper or other source daily (such as *The Wall Street Journal*).
- Attend as many relevant in-person events as your schedule will allow, and review upcoming events to attend, both local and national. (I now post on LinkedIn what events I plan on attending, especially when the event is an FOC event, to help people meet up with me.)

Career Benefits of Maintaining Good Intel

When you actively maintain (there's that word again) your relationships and your intel, you naturally discover job opportunities, often before they are posted publicly. Keeping up with people in the industry connects you to what's really happening, especially who is hiring and when. Want the inside scoop on the next cool security director job? That will happen only if you've kept connected to people who share good intel. Thus, if you've done your work and maintained those relationships, you will not only hear about a job before others do, but you have a better chance of landing that job because you are able to hand over your resume to the person connected to the hiring manager—or, better yet, directly to the hiring manager. Your ability to connect has led to good, actionable intel, which has led to a better job.

Good Intel and Leadership Opportunities

One great way to maintain your intel is by volunteering to take on key roles at key conferences, which in turn can help you gain critical exposure to others in the industry. These key roles open you up to even more opportunities to gather good intel as well. Plus, landing a coveted spot on a discussion panel usually only happens through good intel.

Not only do you have to know about the opening, you need to have a good relationship with those responsible for filling the opening. For instance, your buddy Phil, who is on that panel, was able to get you on that panel because you forged a close relationship with him and because you asked him about any openings—you mined for the intel. How intel works is simple, but you need to put in the effort. When you do, opportunities to lead, influence, and get stuff done always result.

Get your intel from good sources, and do this again and again and again, because the flow of information never stops. When you gather good intel, you strengthen all areas of your reputation and identity within the security world. More importantly, gathering good intel helps strengthen the most important piece of all and the topic for my next and last chapter: relationships.

Chapter 12

Relationships

Without a doubt, relationships are the mecca of the entire security industry. They are *the key* to your success as a security professional. Relationships are the single most important element that will help you find success in the security industry, period—no exceptions. And that is why I want to end this book by talking about relationships.

Relationships are how this industry moves, how your credibility is built, and how you get deals done and gather good intel. Relationships are where the rubber meets the road. The security and law enforcement business is a people business, first and foremost. But, if you disregard relationships, all opportunities for you are null and void. Everyone in security is in the business of ensuring the security and safety of people and assets, and relationships make the difference between success and failure when protecting others. Relationships are key in any people business, but they are especially important in this business.

YES, S.I.R.

Relationships are the be-all and end-all in this industry because your ability to execute, to be pulled into circles, and to discover and obtain golden opportunities hinges on the quality of your industry relationships. Your relationships enable you to connect with the people who can help you achieve your big dream and get stuff done, all while having fun.

I recall one time that I needed to connect a friend on an investigation worth millions of dollars in loss, but he had no clue who to talk with to bypass the formality of typical corporate bureaucracy. Time was critical! John called me. We had not talked in two years, but in FOC years, it was only two days. He was at his wit's end, and thought to call me.

He said, "Chuck, you know everyone, and you're my last resort to get this shipment stopped before it hits the border."

I laughed and said, "Give me what you got!"

He did, and with that information I was able to put in motion years of relationships with the Texas Rangers and other LEO's that ensured that things didn't work out for the bad guys like they'd hoped. I'll just leave it at that. With a couple of 10-95's out of the ordeal, it was a success. That is how relationships work and allow you to get stuff done!

To build and maintain these relationships, you must go beyond the virtual. Sure, Zoom can save you time and money and has become a more acceptable form of communication, but Zoom will never replace real, in-person interfacing and networking. You need to get out and meet these industry contacts in person at networking events and conferences. Meeting in person is still the best way to gather that key intel, execute your strategy, get stuff done, and, of course, develop deeper relationships. With relationships, the physical realm always beats the virtual realm. When you put in the time and effort to build in-person relationships, you no longer have to work the room—the room will come to you, as it now does for me.

RELATIONSHIPS

Remember, good intel and good strategy exist to help you build the most important element, which is good relationships. And it is those relationships that allow you to serve others while you realize your career goals. In other words, good strategy entails, among other things, maintaining a robust LinkedIn profile with a current professional picture that lets you get and stay connected with your relationships. And good intel means everything you post on LinkedIn is current, accurate, and helpful. Of course, good strategy and intel are equally, if not more, important than when you network in person. What I'm saying is that strategy and intel work together to help you build these key, career-changing relationships.

Relationships also help you dream big and fulfill those dreams. Forget the fundamentals of human interactions in this business and you can forget getting ahead and fulfilling your big dreams. I know what I'm talking about here when I say relationships are key. Remember, I'm the guy who procured a C-130 transport aircraft in just two hours, and that only happened because I had invested in those key relationships over time.

Relationships are key because, in a people business like this one, you need to get to know people and keep doing it, regularly. In other words, you need to do maintenance. Yes, maintenance is not easy and can sometimes be exhausting, but it is the key to keeping and deepening key relationships. I don't love answering emails in Dubai at two o'clock in the morning, but when someone who is in a key relationship needs something, I answer.

One of the biggest companies in France hired me based on my reputation alone, a reputation I had built based on—wait for it—a relationship. As I wrote this chapter, a message from the security director from one of the biggest organizations in the United States came across my computer, asking about an upcoming event where I had invited several agencies and organizations, including theirs.

As soon as I saw this message, I paused what I was doing to read his message and follow up on his question with the answers they needed. Keeping relationships top priority means that when you get an urgent message you always answer and follow up—even when you're writing a book!

To sum up, relationships are the key to getting where you want to go in this industry. Through relationships, you will be able to dream big, have fun, and get stuff done, all while gathering the best intel and executing your winning strategy. Relationships are everything, and you need to treat them as such. Forget relationships, and you can forget your dreams, your intel, and your strategy. Forget relationships, and it won't matter how good your strategy or intel is. Forget relationships, and no matter how big you dream or how much fun you try to have, it won't work.

The road to success in this industry travels through key relationships. Remember that, and the security world and all it has to offer are yours for the asking.

Scan to download Friends of Chuck APP

Join Friends of Chuck!

Friends of Chuck (FOC) is a professional security network group that exists for the purposes of networking, locating employment, exchanging business opportunities, discovering new emerging security technologies, and sharing the information that FOC members like you, your companies, and your organization need to know! The FOC is an extensively broad network of 100,000+ Friends of Chuck security professionals across the globe. The common connection is that you know CHUCK and CHUCK knows you!

About Chuck

Charles "CHUCK" Andrews, CPP, CFE, CFCI, ICPS, MSME, has been in the law-enforcement & security industries since 1976, starting as a Law Enforcement Police Explorer in greater Houston, TX. Through his Police Explorer career, he logged 4000 hours of patrol time before his 19th birthday, when he started as a TEXAS Peace Officer.

After completing his Bachelor's Degree in Criminal Justice from the distinguished Sam Houston State University (SHSU), with honors, Chuck studied the criminal justice systems of multiple countries at the Max Plank Institute in Europe. At SHSU, Chuck received first-ever scholarships from the United States Secret Service group and the Sheriff's Association of Texas. Upon his return from Europe, Chuck worked full-time in Law Enforcement in both Colorado and Texas in patrol, crime prevention, and training as a Commander and as Chief of Police. During his law enforcement tenure, Chuck was an advisor or consultant for many different law enforcement/security companies. Chuck also obtained his MSME MBA graduate degree in Security Executive Management from the Univ. of Houston-Downtown, as well as an engineering chemical processing degree from another college. Retiring from law enforcement after being Chief of Police in TEXAS, Chuck began his full-time pursuit of the private security industry, but still maintains today his TX Master Commission Peace Officer status. In the corporate private sector, Chuck served as Chief Security Officer for First Data Corp-TeleCheck and NSS Labs for many years. Currently, Chuck sits on the boards of, or acts in the capacity of advisor, strategist, and/or thought-leader for, security & technology companies around the globe. Chuck is also a board member, owns, and/or is heavily invested in various technology ventures.

During Chuck's career in Law Enforcement & Corporate Security, he uniquely carried out consult work with many different law enforcement agencies and security/LEO companies across the globe, including the well-known reality TV Show – *COPS!* Chuck joined ASIS International ≈33 years ago with the ASIS Denver Mile-High Chapter and then back with the ASIS Houston, TX Chapter where his home membership resides today. Chuck recently sat on the ASIS International Board of Directors with over 33 years of ASIS Intl. volunteer service. ASIS International is the world's oldest and largest security professional society of over 68 years! Chuck is also a member of the ACFE in the Houston ACFE Chapter as a CFE and many other security related organizations like the ACFE, IAFCI, ICPS, InfraGard, PDA & IACP and has held several positions of leadership with certifications.

ABOUT CHUCK

Chuck's career spans many different experiences, trainings, certifications, and disciplines, making him one of the most diverse & sought-after experts in the law enforcement, cyber, fraud, technology & security industries via the many different roles in which he served—Chief of Police, Crime Prevention Director, Chief Security Officer, Asst' Dean – Criminal Justice/Security (Phoenix University), Technology Business Owner, Cyber Company Owner, Trusted Advisor, Cyber Expert, CEO, Adjunct Graduate Professor of Computer Science (Sam Houston State University), as well as SR VP overseeing an $80mm company.

Most notably, Chuck is the founder of ≈100,000 members in 100 countries in the security industry around the globe in his well-known Friends of Chuck (FOC) group. FOC is responsible for helping hundreds find jobs, network, create business deals, and develop relationships in the security & technology industry and all its ancillary parts.

In 2020, Chuck was selected by IFSEC as the World's #1 Security Influencer, globally! From his work with the 1996 Olympic Security Team in Atlanta, GA, to being recognized by the Gov. of Colorado for his work in Crime Prevention, to being named the top Crime Prevention Officer internationally, as well as being awarded OSPA'S Lifetime Achievement Award in 2019, he is recognized as one of the Security industries most influential thought-leaders! Chuck today sits on Boards as Advisor/Strategist to security/cyber/law-enforcement organizations & technology companies. Chuck is regularly sought after as an SME in matters of Security (Cyber & Physical) & Law Enforcement in multiple verticals, as Board Member, Advisor & Strategist in emerging security technologies, cyber, software and other cool stuff on the planet.

YES, S.I.R.

In short, Chuck is one of the top sought-after thought-leaders of the security & technology industries!

www.friendsofchuck.com
https://www.linkedin.com/in/charlesandrewscpp/
https://www.facebook.com/friendsofchuckFOC/
https://twitter.com/FOChuck
Chuck can be contacted at: charles.andrews.cso@gmail.com

What My Friends Say (cont.)

"Chuck and I have a lot in common, from the early days of considering a police career to the later days volunteering in the security community. We both have a passion for connecting others and helping some understand the benefits of what being connected can do to your career. Good job, Chuck."

– Ray O'Hara, CPP

"Many people know Chuck Andrews as "The Hat." And The Hat is indeed instantly recognizable and compelling. But underneath that ten-gallon topper is a powerful brain that has mastered security and its intricacies and a mighty heart that thrives on helping colleagues advance their careers. That's my definition of an influencer."

– Michael Gips, CPP, CSyP
Principal, Global Insights in Professional Security,
IFSEC 2022 #1 Most Influential Person in Security,
thought leadership category

"I retired from the United States Army ten years ago and have known Chuck for almost my entire civilian life. We met at an ASIS International event, and I immediately knew he was someone I wanted to emulate. He took me under his wing and coached me as a chapter vice chair and assistant regional vice president. I eventually asked Chuck to do

YES, S.I.R.

something he takes very seriously—and today I am proud to refer to him as my mentor, as well as a wonderful friend!"

— **Eric Kready, CPP**
Area Security Manager, SAP/Director,
ASIS International Professional Certification Board (#3)

"Networking and relationships are somewhat simple to start. Chuck Andrews goes way beyond the start of any relationship he starts and offers long-term continuity and friendship. Not only do you feel the love of this big Texas man, you know it's genuine and sincere. Chuck continues to make himself available and always willing to share with you even though his calendar may be full. Chuck is a plow and seed planter who never runs out of seeds. The security, public safety, and military sectors are lucky to have him and 'Friends of Chuck.'"

— **Mark Eklund**
Vice President of Security Solutions,
Xandar Kardian, Inc.

"Chuck, thank you for your guidance and support as I try to navigate the world of security influencers. I appreciate your time guiding me about my strategy and goals and always focusing on my passion. You have supported and believed in me when I was tired, unfocused, and downright giving up. I have learned a great deal from you, my friend. There is no one in the security industry, and for that matter, any industry that can write this kind of book. You are a great mentor, a confidant, and a friend to me. Your energy is contagious, and your love for what you do shines everywhere, and with every person you reach. I look forward to YES, S.I.R. and can't wait to continue to learn."

— **Carlos Francisco, CPP**
"The Corporate Security Translator"

WHAT MY FRIENDS SAY (CONT.)

"I have been influenced by Chuck since 2010 as classmates and have leaned on him every step of my professional career as a true mentor. He has shown me the power of collaboration and networking as well as the importance of brand recognition. He is always forward-leaning into new technologies as well as cutting-edge policy improvements."

– **William Clement**
Security Manager, South Texas College of Law,
Houston and Master of Security Management

"Chuck has the energy and charisma of ten people, the dedication of a Ranger, and the directional fortitude of a crusader. Known for his personal branding and his complete focus on enhancing the security profession, Chuck is to the security and law enforcement communities as a herald of the new breed of professionals."

– **Joe McDonald, CPP, PSP, CMAS**
IFPO – Board of Advisors

"Chuck gets real results. He has established an effective national and international network of professionals who supply exceptional products and services. He does it the old-fashioned way—through hard work and becoming their brand ambassador. He is a great advocate on your company's behalf."

– **Steven Doran, CHPP**
Internationally recognized expert and consultant

"If there's one thing I've learned from Chuck, it's the importance of creating relationships. In 1990, at the age of fourteen, I showed up at the back door of the police department to join the Explorer program and was greeted by Chuck, my advisor. Chuck soon had me involved in building haunted houses, wearing a McGruff the Crime Dog suit, and forging relationships with the community. The skill set I learned from his years of mentorship helped shape my thirty-year career in law enforcement."

– Vinnie Montez
Commander at Colorado Sheriff's Department

"Chuck Andrews is a dynamic influencer who has mastered the art of relationships and connectivity. Our candid conversations about society's challenges with bridging gaps between law enforcement, private security, and the minority community have always been of value. I appreciate the inspiration he infectiously delivers, and I am proud to refer to him as a friend."

– Joe Moon, ARVP
ASIS Region 3-C (NTX Chapter)

"Relationships are everything, regardless of what you do, where you live, what position/title you have or what stage of life you are in. Chuck realized that a long time ago, and his efforts via the Friends of Chuck (FOC) epitomizes the value and lasting importance of relationships. Countless people, myself included, have benefited from this extended family of friends and professionals who will lend an ear, a hand. or a thought whenever asked."

– Richard E. Widup, Jr, CPP, CFE
Global Enterprise Leader/President and Founder,
The Widup Group, LLC

"Few people I know understand the importance of a broad network like Chuck Andrews. He is the epitome of the servant leader, giving deeply of himself to those in his chosen profession specifically, and society in general. Our world is a better place because of him."

– Mike Mallon
Storyteller Promotions

WHAT MY FRIENDS SAY (CONT.)

"Besides being one of the most recognizable figures in the security industry, Chuck has filled a niche in the security community as a professional connector. Due to his unparalleled network (FOC), Chuck is able to connect security practitioners around the world who are looking for specific skills, products, and services. It has been my absolute pleasure to count Chuck as a trusted peer and valued friend. Thank you, Chuck, for connecting the security community."

– **Scott Lowther**
Park Way Properties, LLC

"Chuck Andrews is an accomplished security professional and consummate networker. He's also incredibly supportive of others in the security industry. For instance, my company was considering entering the industry and I was doing some market research. I found Chuck on LinkedIn, had a cup of coffee with him, and not only did he have complete command of the market but he introduced me to some key people and helped recruit the individual who led that line of business and grew it into an important aspect of our firm. At the time, he didn't know me from Adam: just a giving guy willing to support others in the industry."

– **Dave Jacobs**
Principal, TechKnowledge Consulting

"Over the past two decades, I've had the pleasure of working with Chuck in various settings, and getting to know him and his lovely family in both a professional and personal capacity. Just a glance at both of our professional resumes reveal the many times our paths have crossed over the years, and that is not by accident. Together we have navigated troubled waters while also sharing many more successes, and through it all, his positive outlook never waivered; regardless of the challenges, he was always the consummate professional and truly a gentleman.

YES, S.I.R.

Chuck is truly a 'maven' and never passes up an opportunity to lend a helping hand. He is one of the most dedicated and qualified security professionals that I have ever had the pleasure of knowing and working with. There are few people in this industry that I would hire again without hesitation. Being a successful leader in Security requires developing trust; and this profession is built on strong relationships; FOC is living proof of trusted relationships. His is a friendship that I truly cherish and will continue to withstand the test of time."

– **Linda Florence, Ph.D., CPP**
Industrial/Organizational Psychologist
The Florence Group
Violence Prevention Consultant,
Career Security Professional